Visual Balance + Simplicity + Storytelling = Process-bility

BPMN Made Easy with Process-bility
Effective Visual Communication for Digital Transformation

Published by Process-bility Press

www.process-bility.com

ISBN: 978-1-7638978-0-9

To Mauren,

Thank you for being my rock behind the scenes, looking after everything, so I could focus on writing. You've given me the time and space to burn the midnight oil and bring this project to life. I couldn't have done it without your love, patience, and unwavering support.

With all my heart,

Kevin

2025 May 06

Foreword

With over 20 years of global experience, working for major companies like Siemens and SAP in 7 countries, I've had the privilege of solving some of the most complex business challenges by aligning People, Process, Applications, and Data. Throughout my career as a former CIO, CFO, Head of Operational Excellence, and Head of Service Business, I've seen first-hand how important process modeling is in every business stage, from start-ups to acquisitions.

In the world of process modeling, complexity often overshadows clarity, leaving many professionals—both beginners and experts—struggling to communicate their ideas effectively. That's where Kevin Tan steps in, with an unwavering passion to demystify the art of Business Process Model and Notation (BPMN). Through his years of experience and dedication, Kevin has developed a fresh approach that blends simplicity, storytelling, and visual balance, making process modeling accessible to everyone.

As I turned the pages of BPMN Made Easy with Process-bility, I found myself not just reading but discovering. Kevin's insights opened my eyes to the nuances of process modeling in a way I had never imagined. What initially seemed like a straightforward guide quickly became an ocean of value waiting to be unlocked. Every chapter revealed new dimensions of BPMN, and I came to realise just how much potential was hidden beneath the surface of what I once thought was a rigid, complex framework.

The more I delved into the principles of Storytelling and Visual Balance, the more I appreciated the depth and power of Kevin's approach. His ability to break down complicated concepts into digestible, actionable steps made it clear that BPMN is not just a tool for experts, but a language anyone can master. And as I continued to explore the pages, I felt confident that I was gaining skills that could drive real transformation in how processes are modeled, understood, and communicated.

This book, BPMN Made Easy with Process-bility, is more than a guide; it's an invitation to explore a new way of thinking about processes. Whether you're just starting out or are a seasoned analyst, Kevin's insights will resonate with you. His desire to share his expertise and help others overcome the challenges of process modeling is evident on every page. Kevin's belief that process models should not only function but also inspire collaboration and innovation is at the heart of this work.

Dive in, and you'll find that this book is not just for experts but for anyone who wants to unlock the true potential of BPMN. With Kevin as your guide, complexity gives way to clarity, and the art of process modeling becomes an empowering tool for everyone.

Gia Thi Nguyen

VP, SAP Signavio, Market Impact APJ & GC
Singapore, Oct 2024

Dear Reader,

Welcome to the world of Process-bility! Thank you for picking up BPMN Made Easy with Process-bility. This is more than just a book — it's a journey to simplify BPMN, drive real business impact, and transform the way we model processes.

With Process-bility, you're not just mastering BPMN—you're unlocking a pathway to better insights, reduced complexity, and real business results. I hope this book inspires you to break down silos, improve collaboration, and create lasting value in your organization.

I'd love to hear how Process-bility is shaping your work!
Feel free to connect anytime.

Let's make BPMN easy!

Kevin Tan
Author, BPMN Made Easy with Process-bility

Preface

In the world of business process modeling, it's easy to get lost in complex notations, gateways, and jargon. For years, I saw professionals—new and experienced—struggle to create models that worked and could be easily understood across an organization. That struggle inspired the creation of Process-bility.

Born from two decades of hands-on experience, Process-bility makes BPMN easier, clearer, and more intuitive—without losing its power. It focuses on three core principles: Storytelling, Visual Balance, and Simplicity, ensuring models not only function but also tell engaging stories that everyone—from CEOs to frontline employees—can follow.

Along the way, I learned that process modeling isn't just about technical accuracy— it's about clarity, growth, and collaboration. When models are easy to understand, they spark conversations, align teams, and encourage continuous improvement.

This book isn't just about building better models—it's about communicating better, working together, and enjoying the process. Whether you're a seasoned business analyst or just starting your journey, I hope Process-bility provides the tools, insights, and confidence to unlock your potential. Let's turn complexity into clarity and make BPMN easy with Process-bility.

Kevin Tan
www.Process-bility.com

**To inspire people to rediscover their growth mindset,
be nice to each other,
and have fun.**

CONTENTS

If you can’t model your business process, you don’t understand it well enough.

Chapter 1

Introduction

- From Complexity to Clarity
- Why This Book?
- The Problem with BPMN
- What You'll Learn
- Why Process-bility Matters?
- Closing Thoughts

From Complexity to Clarity

Have you ever tried to navigate the world of Business Process Model and Notation (BPMN), only to find yourself tangled in its complexities? The notations, Gateways, Flow Lines — it can all feel overwhelming, even for seasoned professionals. But BPMN is critical. In today's fast-paced, digital-first world, the ability to model and truly understand your business processes is key to successful digital transformation and continuous improvement.

Process-bility strips away the complexity from BPMN, replacing it with clarity, simplicity, and storytelling. At its core, process modeling is more than just diagrams and symbols — it's about telling a story. A visual story about your organization, your people, and how they collaborate to achieve meaningful business outcomes.

Why This Book?

With over three decades of experience in business process modeling, I've encountered it all — the good, the bad, and the unnecessarily complex. Through these experiences, I developed a set of principles called Process-bility. This methodology breaks down BPMN into its most essential components, focusing on three core principles:

1 Visual Balance

A well-designed BPMN model should be visually balanced. That means no messy cross-over lines, no cluttered activities, and an easy-to-follow flow from start to finish. A visually balanced model is not just more aesthetically pleasing—it's easier to understand, which leads to better decisions.

2 Simplicity

BPMN can quickly get bogged down in details and notations that add more complexity than value. The best BPMN models are simple but not simpler. They contain just enough information to convey the process clearly, without overwhelming the reader.

3 Storytelling

At its core, every BPMN model should tell a story that anyone—from a CEO to an entry-level analyst—can understand. The focus is not on showing off how many notations you know, but on creating a clear, engaging narrative that communicates the process effectively.

The Problem with BPMN

BPMN is undoubtedly a powerful notation standard, capable of supporting both events and activities, as well as business rules and decision flows. However, its very strength can also be its downfall. With hundred of notations to choose from, and no established best practices, many process modelers fall into the trap of focusing too heavily on technical complexity. Often, models are driven by the need to represent every detail, rather than focusing on business clarity.

This technical bias leads to over-complicated models where many business rules and decision flows are treated as processes themselves. The result is a model that, while technically accurate, is confusing and overwhelming for all stakeholders. Instead of providing clarity, these models muddy the waters by including unnecessary detail, making it harder for teams to understand the processes.

When process modelers attempt to represent every nuance, decision, and rule in a BPMN diagram, they often lose sight of the bigger picture. Without a clear framework, it becomes easy to overcomplicate the model. The true purpose of BPMN—capturing and communicating a clear, digestible narrative of how a business operates—gets lost in the details.

Process-bility offers a superior approach. By focusing on Visual Balance, Simplicity and Storytelling, it ensures that BPMN models serve their real purpose: communicating not just how a process works, but why it matters to the business.

What You'll Learn

In this book, you'll learn how to approach BPMN with the same mindset I developed through years of trial and error. You'll discover how to:

- Use Visual Balance to create BPMN models that are easy to read and understand.
- Tell a story with clarity using BPMN models - with minimum notations.

- Apply the principles of Storytelling to cut through the noise and deliver BPMN models that drive real, actionable insights.

We'll also look at common BPMN mistakes and how to avoid them, share real-world examples of BPMN models done right (and wrong), and give you practical tips on applying Process-bility in your day-to-day BPMN modeling.

Whether you're new to BPMN or a seasoned professional, this book will show you how to take your BPMN process modeling to the next level.

> **"Absorb what is useful, discard what is not, add what is specifically your own.**
>
> BRUCE LEE

Why Process-bility Matters?

The business landscape is changing at an unprecedented pace. Digital transformation and continuous improvement are no longer a buzzword —it's a necessity for survival. And as businesses pivot toward automation, artificial intelligence, and data-driven decision-making, having a clear understanding of your processes becomes even more critical.

Process-bility helps you see the bigger picture. It's not just about creating a diagram that looks good — it's about fostering collaboration, breaking down silos, and driving meaningful change within your organization.

As we dive into the world of BPMN through the lens of Process-bility, you'll see how a well-modeled process can unlock new efficiencies, improve communication, and ultimately, help your business thrive.

Fg1.1 The future is now

Closing Thoughts

Process is the manifestation of activities when knowledge is applied to achieve a desired business outcome. Think of BPMN as the X-ray of your business processes. Just like an X-ray helps doctors diagnose a problem, a well-modeled process helps you diagnose inefficiencies, identify opportunities, and ensure that your business is operating at its full potential.

So let's get started on this journey together — from complexity to clarity, from confusion to understanding. BPMN made easy with Process-bility.

> **Process is the manifestation of activities when knowledge is applied to achieve a desired business outcome.**

Chapter 2

The foundation of BPMN

- What is BPMN?
- The Key Elements of BPMN
- Why Most People Get It Wrong?
- Get Ready With Process-bility
- Closing Thoughts

What is BPMN?

Business Process Model and Notation (BPMN) is an international standard developed to visually represent business processes in a way that can be easily understood across various roles—from stakeholders to developers. Initially introduced by the Business Process Management Initiative (BPMI) in 2004, it was later adopted by the Object Management Group (OMG) in 2005. The goal was to bridge the gap between business and IT by providing a universal language for process modeling, and BPMN quickly became the go-to standard. The release of BPMN 2.0 brought enhancements, including expanded event and data flow capabilities, making it even more robust for modern business process management needs.

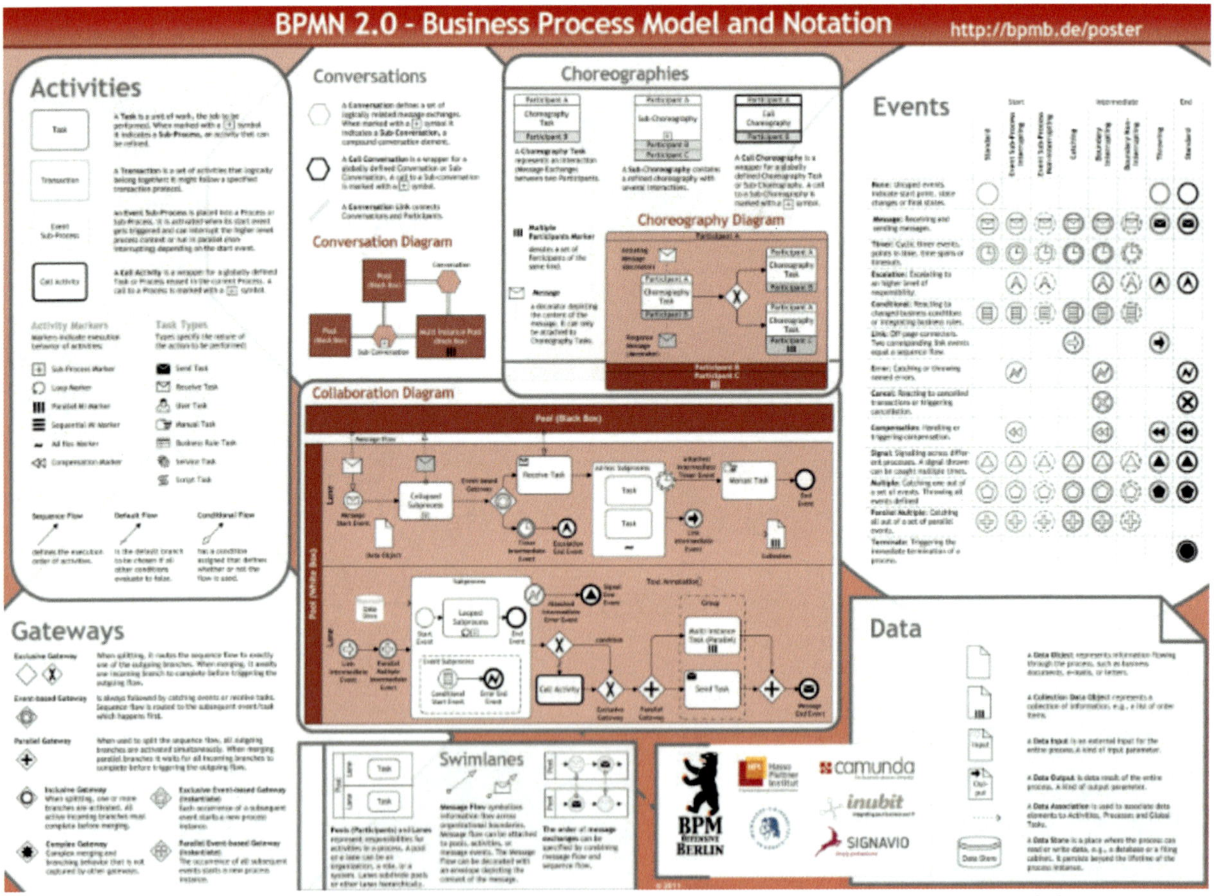

Fg2.1 BPMN 2.0 Notations

However, while BPMN 2.0 provides a rich, comprehensive framework, its official OMG documentation is extremely technical, more like a programming manual than a user-friendly guide. The dense, detail-oriented nature of the standard has made it challenging for newcomers to approach. Consequently, many turn to online resources to learn BPMN, often relying on tutorials or summaries that lack depth and accuracy. This has led to widespread misinterpretations and inconsistent practices, resulting in overcomplicated or ineffective models. Instead of fostering clarity, BPMN* often ends up misunderstood and misused—its intended simplicity lost in translation.

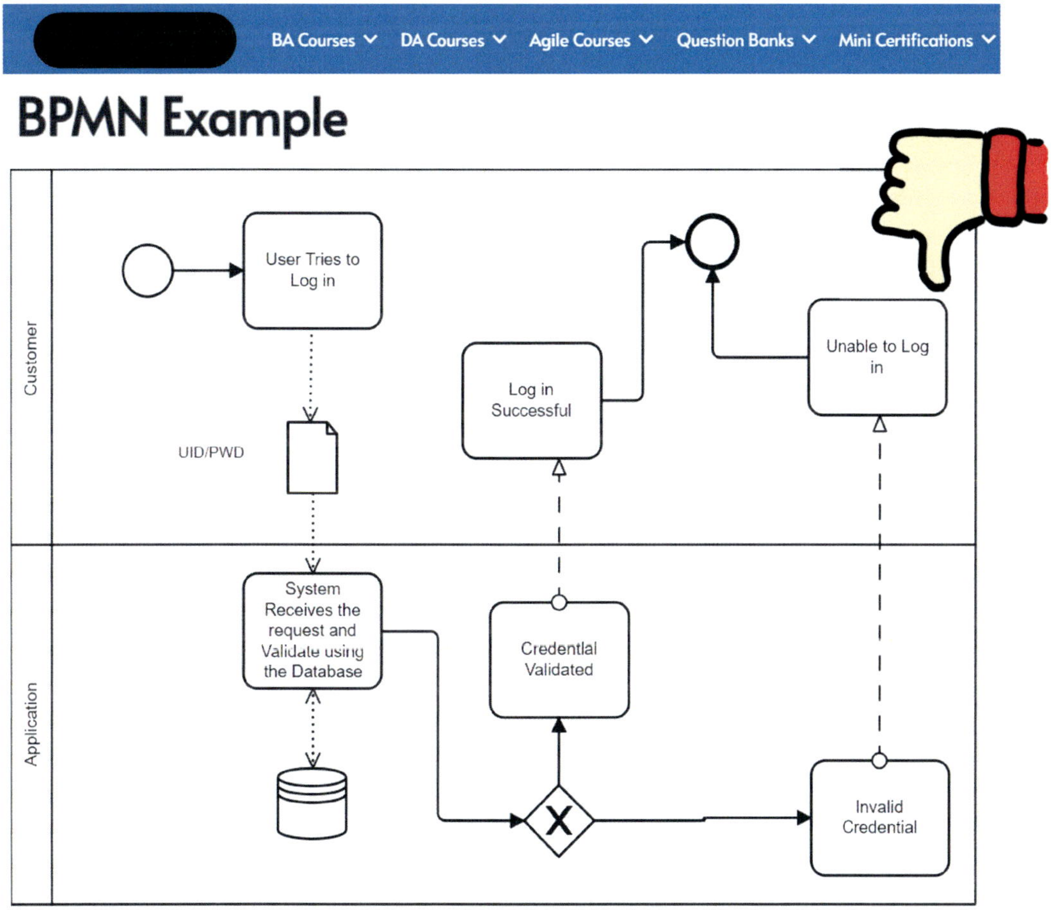

Fg2.2 Poor quality BPMN online resources to learn BPMN

BPMN* specifically refers to Collaboration Diagrams within the BPMN 2.0 standard. While BPMN 2.0 defines four types of diagrams—Process, Collaboration, Choreography, and Conversation—our focus in this eBook is exclusively on Collaboration Diagrams, where inter-team interactions and message flows are clearly visualised.

The Key Elements of BPMN

At its core, BPMN revolves around five key elements:

Event, Activity, Gateway, Swimlane, and Connecting Flow.

These are the fundamental "building blocks" of process models, with each playing a critical role in representing different aspects of a business process. The remaining notations are essentially variations of these core elements, providing additional detail or specificity, but they all function based on the same foundational behaviors. By mastering these five key elements, you can navigate BPMN without being overwhelmed by its many notations.

Fg2.3 BPMN Key Elements

Event

An Event represents something that occurs during a process. It can trigger the start of an activity, mark its end, or serve as a checkpoint during the process. Events are essential for understanding how the process flows and how it responds to various triggers or changes.

Examples of Events include Start Event, Message Start Event, End Event, and Intermediate Timer Event, among others.

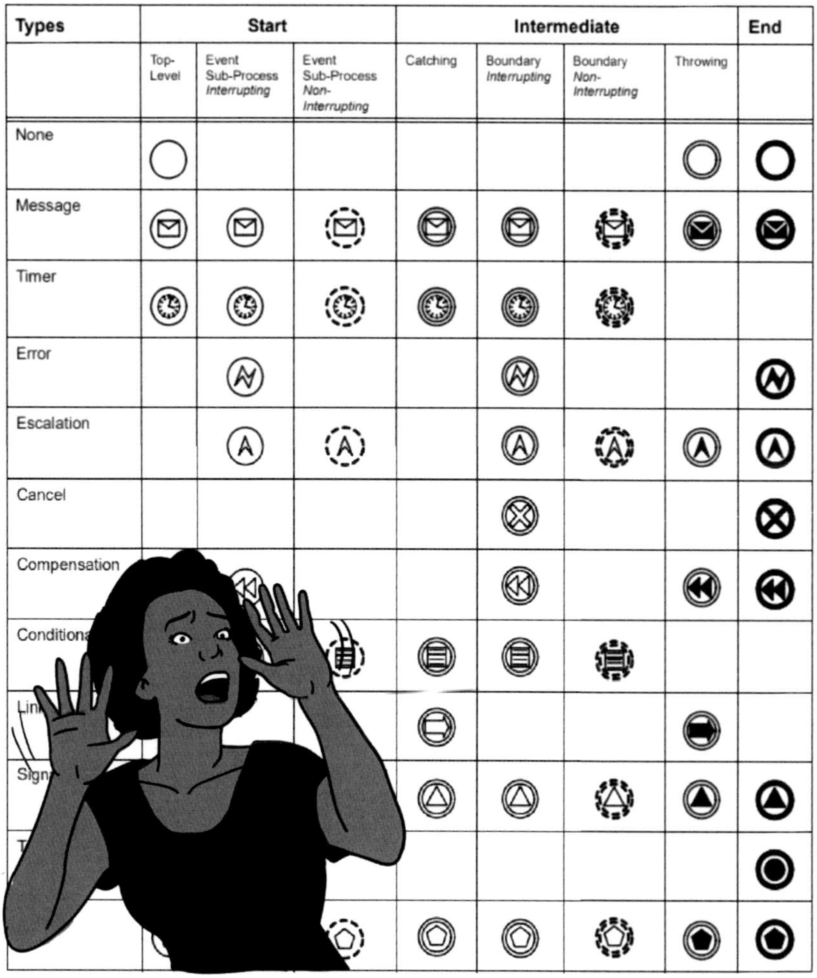

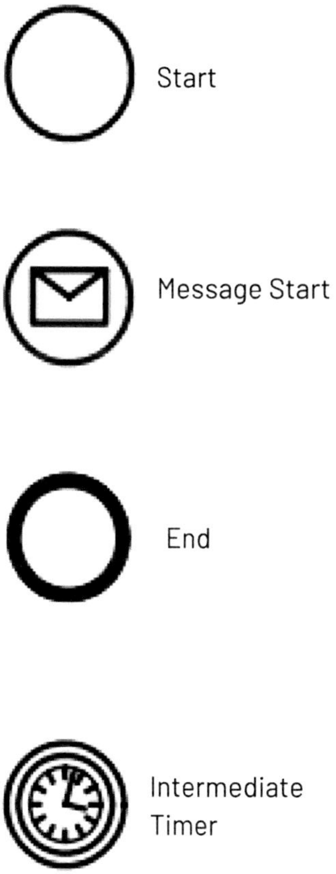

Fg2.4 BPMN Notation - Events - Don't worry, you only need to understand less than 10 BPMN notations to be able to master BPMN modelling

Activity

An Activity represents tasks or work that is being performed. These can range from manual tasks (like destroying duplicate hardcopies) to automated tasks (like validating credit status). In BPMN, activities can be further categorised into Tasks, Sub-processes, and Call Activities, but their core purpose remains the same—they are the actions that move the process forward.

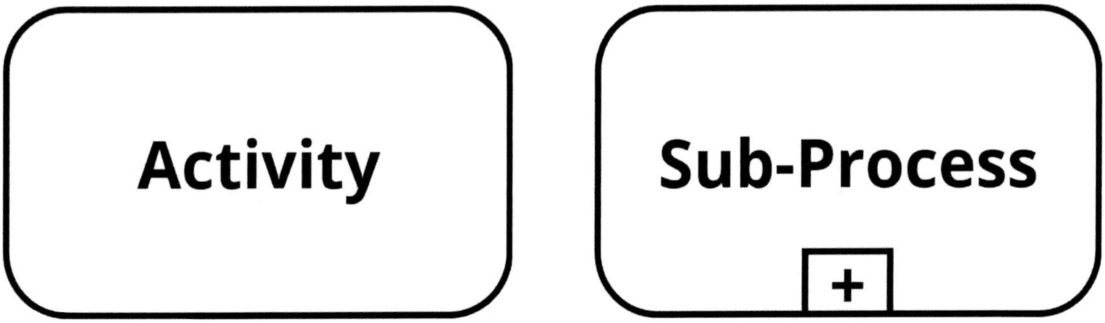

Fg2.5 BPMN Notation - Activity and Sub-Process

Gateway

Gateways control the flow of a process, acting as routing points where the flow can either split into different paths or converge back into one. They ensure that processes adapt to various scenarios, but they are often misunderstood and overused.

The most commonly used Gateways are the Exclusive Gateway and Parallel Gateway, while the Inclusive Gateway and Event-based Gateway are less frequently used.

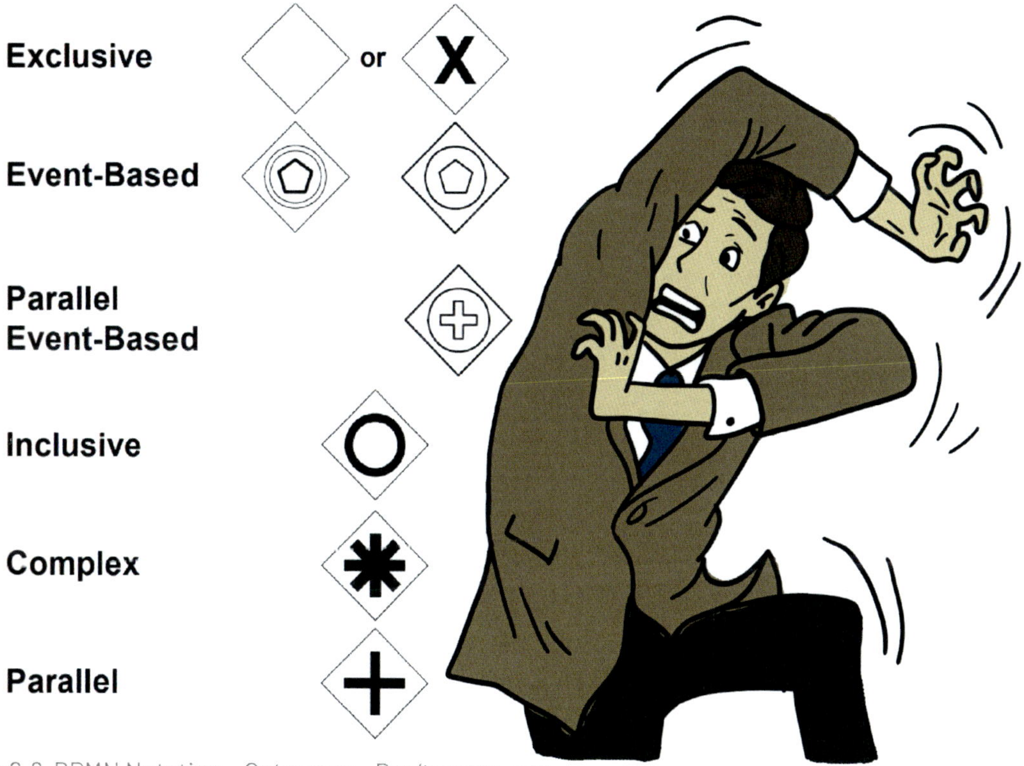

Fg2.6 BPMN Notation - Gateways - Don't worry, you only need to understand 4 Gateways to model most of the processes.

Swimlane

Swimlanes represent the roles, teams, or systems involved in the process. They visually organise the process by assigning tasks to specific roles, making it easier for stakeholders to understand their responsibilities and contributions within the process.

Here are the variations of Swimlane:

- Pool
- External Pool

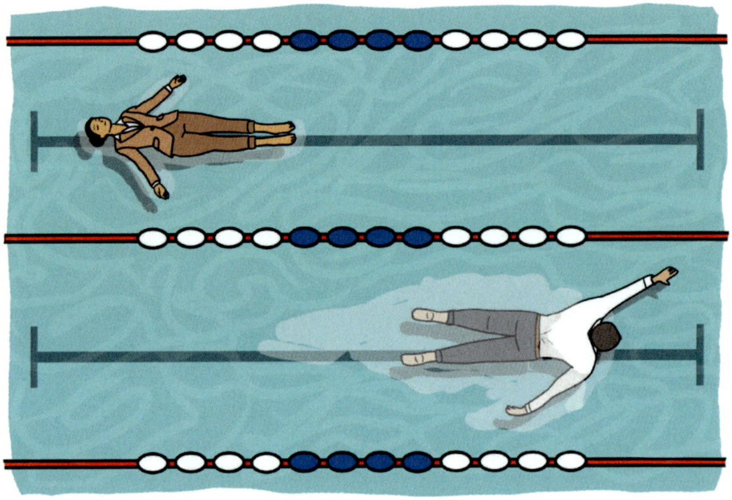

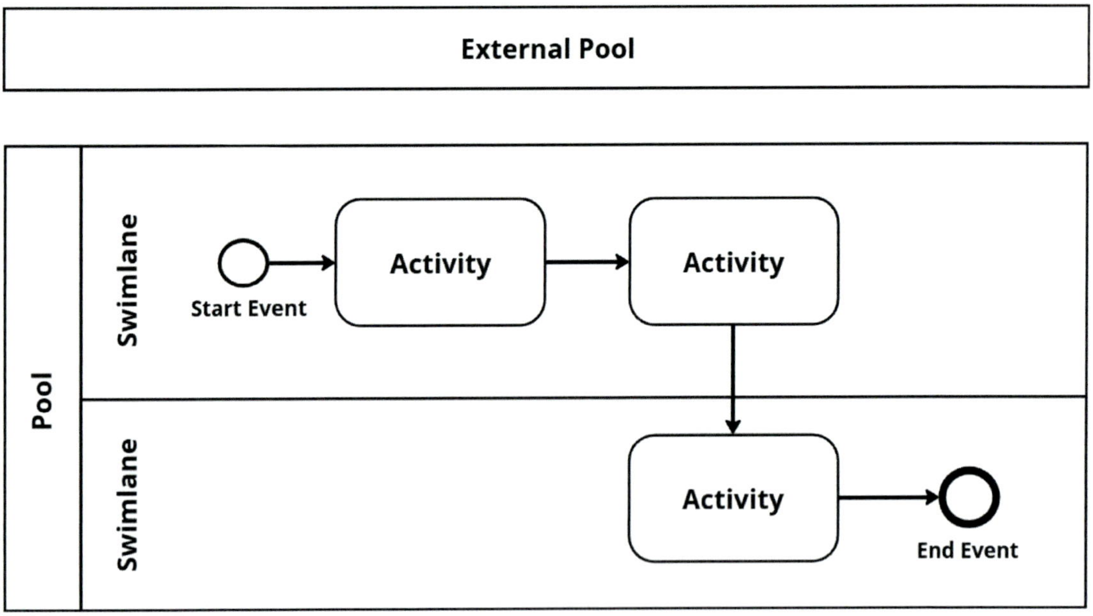

Fg2.7 BPMN Notation - Swimlane, Pool and External Pool

Connecting Flow

Connecting Flows are the lines that link Activities, Gateways, Events and Artifacts. They depict how the process transitions from one task to the next, ensuring that the flow of the process remains clear and easy to follow.

Here are the variations of Connecting Flows:

- Sequence Flow: Connects Activities, indicating the order of Activities in the process.
- Message Flow: Represents communication between different two Pools.
- Association Flow: Links an object, such as an attachment or note, to an Activity.

Kind:	Depiction:	Specific Depiction Resolution:	
		bpmnElement:	BPMNShape Attributes:
Sequence Flow	Label	SequenceFlow where default is false and conditionExpression is unspecified.	None
Conditional Sequence Flow	Label	SequenceFlow where default is false and conditionExpression is specified (exception when source is a Gateway).	None
Default Sequence Flow	Label	SequenceFlow where default is true and conditionExpression is unspecified.	None
Message Flow	Label	MessageFlow	messageVisibleKind is unspecified.
Initiating Message Flow with Decorator	Label	MessageFlow	messa[illegible]isibl[illegible] initi[illegible]
Non-Initiating Message Flow with Decorator	Label	MessageFlow	mess[illegible]sibleKin[illegible] non-ini[illegible]
Association	Label	Association where associationDirection is none.	[illegible]
Directional Association	Label	Association where associationDirection is one.	[illegible]

Fg2.8 BPMN Notation - Connecting Flow - Don't worry, you only need to understand 3 to be able to master BPMN modelling

Why Most People Get It Wrong?

Despite having a standardised notation, BPMN is often misused or overcomplicated. Many people approach BPMN as a technical skill rather than a storytelling tool, and that's where the confusion starts.

Here are the most common mistakes:

Overcomplicating the Model

It's tempting to include every possible detail in your BPMN model. You might want to show every decision, exception, and path. But just because you can, doesn't mean you should. Overcomplicated models confuse the very people who need to understand them—your stakeholders.

Misuse of Gateways

A significant issue in BPMN modeling is the lack of understanding of Gateways. Many modelers misapply Gateways, resulting in confusion and misinterpretation of the business scenarios and process flow.

Improper merging or branching of Gateways is another common mistake, often disrupting the clarity of the model. Additionally, BPMN's flexibility sometimes leads modelers to either overuse or underuse Gateways, further complicating the diagram.

The key isn't just knowing how to place a gateway but understanding when and why it's necessary to accurately depict process flows.

Focusing on Notation, Not Communication

Too often, BPMN modelers focus on getting the notation right without considering whether the model actually communicates the process clearly. BPMN is a visual language meant for communication, not a technical diagram reserved for the elite few.

Lack of understanding of Sequence Flow and Message Flow

These two notations serve distinct purposes, and confusion between them can significantly impact the accuracy and clarity of the model.

Misuse or Lack of Sub-Processes

Another common mistake is failing to use sub-processes effectively to simplify and structure your model. Sensible grouping of related activities into sub-processes not only declutters your diagram but also helps cater to different audiences—those who prefer high-level overviews, those interested in detailed steps, and even those focusing on technical specifics.

Without proper use of sub-processes, models become overwhelming and difficult to navigate, limiting their effectiveness.

Fg2.9 The tangling kitten

Get Ready With Process-bility

It's clear that mastering the notation alone isn't enough. A model might follow BPMN rules but still fail to communicate effectively. Rather than just focusing on the technical aspects of BPMN, Process-bility transforms BPMN into a communication tool that bridges the gap between complex processes and clear understanding.

At its core, Process-bility is built on three guiding principles:

Visual Balance

Visual Balance ensures the model is easy on the eyes and easy on the mind. Proper alignment of elements, consistent spacing, and clear organization help the reader grasp the process at a glance. It's not just about aesthetics—it's about making the process model engaging and digestible.

Simplicity

Simplicity doesn't mean leaving out important details. It means showing just enough information to tell the story without overwhelming the reader. In Process-bility, we focus on "Less is more." We avoid unnecessary notations and only include elements that help tell the story. These principles are designed to ensure that your BPMN models are not just technically accurate, but easy to understand and impactful for all stakeholders.

Storytelling

Every process model tells a story - with a beginning, middle, and end. The goal of BPMN is to ensure that this story is told clearly so everyone can understand it. In Process-bility, we focus on ensuring the process flow is easy to follow and that each task, event, and decision fits within a coherent narrative.

> **Simplicity is the ultimate Sophistication**
>
> Leonardo da Vinci

Closing Thoughts

Understanding the foundations of BPMN is crucial, but it's just the beginning. Once you grasp the basic elements of BPMN—Activities, Events, Gateways,
Swimlanes, and Connecting Flows—you're ready to take the next step: transforming your models into clear, engaging, and visually balanced stories.

Process-bility is here to guide you through that transformation. In the next chapters, we'll dive deeper into each of these elements, explore common pitfalls, and show you how to apply Process-bility principles to make BPMN models that are both easy to create and easy to understand.

With Process-bility, you'll not only create better models but also help your team and stakeholders understand the process more effectively.

> **There is a science in the art of business process modeling.**

Chapter 3

Gateways With Process-bility

- Are you ready for the Gateway?
- Exclusive Branching Gateway
- Exclusive Merging Gateway
- Parallel Branching Gateway
- Parallel Merging Gateway
- Inclusive Branching Gateway
- Inclusive Merging Gateway
- Event-Based Gateway
- Closing Thoughts

Are you ready for the Gateway?

Gateways are one of the most misunderstood elements in BPMN. While they are essential for both controlling the flow and merging multiple paths in a process, their purpose is often confused, leading to overly complex or inaccurate models. In Process-bility, we aim to simplify the use of gateways, ensuring they serve their dual purpose effectively—whether it's routing the flow based on decisions made elsewhere or merging incoming flows depending on the gateway type.

In this chapter, we'll explore four key types of gateways:

Exclusive Gateway,
Parallel Gateway,
Inclusive Gateway and
Event-based Gateway.

Through clear explanations and analogies, we'll help you understand how to use **each one effectively.**

Fg3.1 The Gateway Heros

Exclusive Branching Gateway 

The Exclusive Branching Gateway (also known as the XOR Gateway) is the most commonly used Gateways in BPMN model. Its role is to route the flow of the process to one of several possible paths based on a decision that has already been made. Despite its simplicity, the Exclusive Gateway is often misused or misunderstood.

What It Does:

Routes the flow based on decisions made in prior task(s) or activitiy(s). Chooses only one path to continue the process. In the example, the Exclusive Branching Gateway - Min. requirement met will route the flow to Activity C if the Min. requirement was met, it will route the flow to Activity D if the Min. requirement was not met, based on the decision made in either Activity A or Activity B.

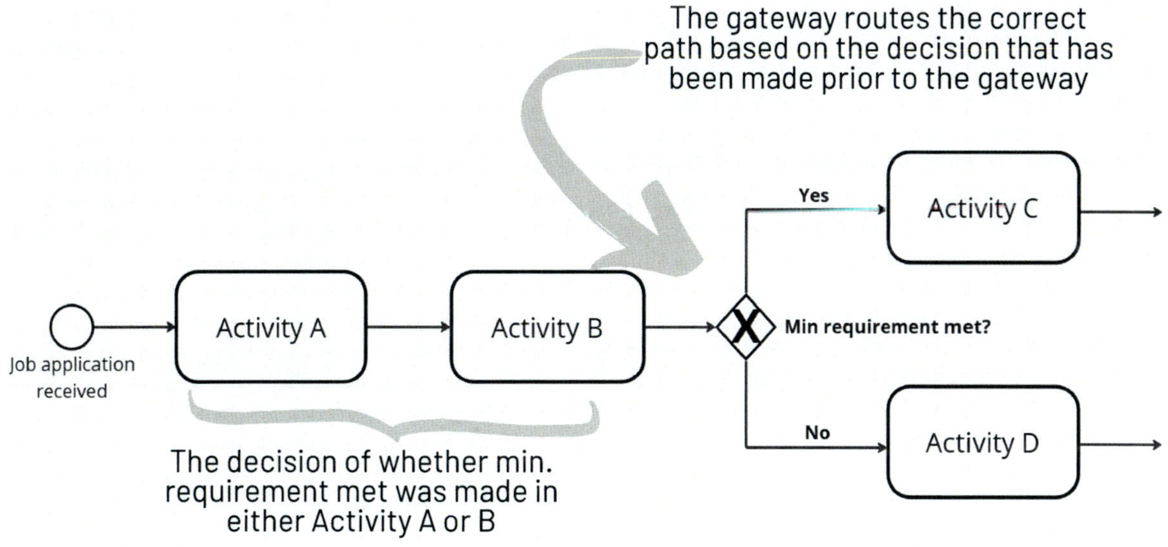

Fg3.2 BPMN Exclusive Branching sample

What It Doesn't Do:

The Exclusive Gateway does not make decisions—this is a common misconception. The decision is always made in the task or activity prior to the gateway. The gateway simply routes the flow based on that decision.

The Toilet Signage Analogy

Think of the Exclusive Branching Gateway like the signage outside public restrooms—one for men, one for women, and others. The decision of whether to enter the male or female restroom isn't made by the signage itself; that decision was already determined by the person before they approached the restroom. The signage simply directs you to the appropriate restroom based on a decision made long before you arrived at that point.

This is exactly how the Exclusive Gateway operates. The decision has already been made in the task or activity before the gateway, and the gateway is simply there to route the process flow to the correct path—just like restroom signage directing you to the right door based on what you already decided.

Fg3.3 Exclusive Gateway Analogy

Exclusive Merging Gateway

The Exclusive Merging Gateway routes two or more incoming flows into a single outgoing flow. It typically reflects an earlier decision made by an Exclusive Branching Gateway (XOR gateway), where one of several possible paths was selected based on specific conditions.

When to Use an Exclusive Merging Gateway:

- Converging Paths: When there are multiple possible branches in a process (e.g., paths that result from an exclusive branching decision), and you want to bring them back together into a single flow.
- Simplifying Flow Control: If you have different tasks or processes that lead to the same final step, the exclusive merging gateway can be used to ensure that the process continues through a single path, regardless of which of the multiple paths was taken previously.

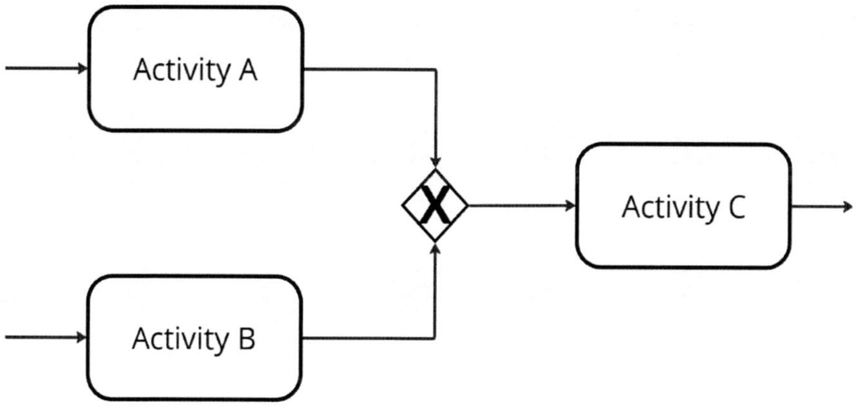

Fg3.4 BPMN Exclusive Merging sample

Exclusive Merging Gateway is not recommended in Process-bility

The exclusive merging gateway is not recommended because it tends to add unnecessary complexity to process models without offering significant clarity.

Here are the key reasons why it is discouraged:

- **Redundant Notation:** The Exclusive Merging Gateway is often used when there are multiple incoming flows, with the intent of merging them back into a single path. However, this is something that can naturally occur without the need for an explicit merging gateway. BPMN models allow for multiple incoming flows to merge without the need for a merging gateway, making the gateway redundant in most cases.

- **Visual Clutter:** Adding gateways where they aren't needed contributes to visual clutter. Since Process-bility emphasises Simplicity and Visual Balance, including unnecessary gateways detracts from the model's clarity and makes it harder for stakeholders to understand the process at a glance.

- **Misleading Representation:** Using an Exclusive Merging Gateway might imply that a decision is being made at that point, which is not the case. The decision should have already been made earlier in the process (before the exclusive branching gateway), and the merging gateway is simply consolidating the paths. Without proper understanding, this can confuse readers into thinking there is an additional decision happening at the merge.

- Encouraging Simplicity: Process-bility advocates for the "less is more" approach. Instead of overloading the model with gateways for every merging point, simply connecting the flows without an explicit merging gateway simplifies the model and maintains the narrative clarity.

By avoiding Exclusive Merging Gateways, Process-bility encourages streamlined, easy-to-understand models that enhance communication rather than complicate it.

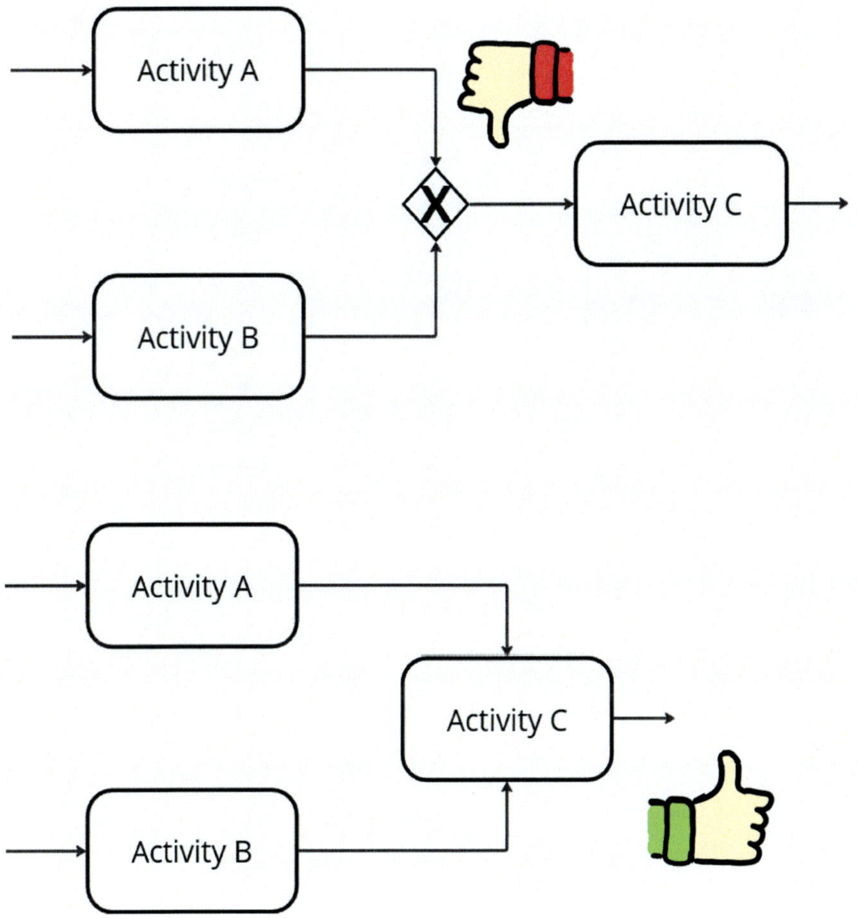

Fg3.5 BPMN Exclusive Merging - not recommended by Process-bility.

Parallel Branching Gateway

The Parallel Branching Gateway is used to manage activities that can occur simultaneously. It doesn't evaluate conditions or make decisions but simply activates multiple branches of a flow.

When the process flow reaches a parallel Branching gateway, all outgoing flows are triggered at once. This means several tasks can begin simultaneously, allowing different parts of the process to progress in parallel.

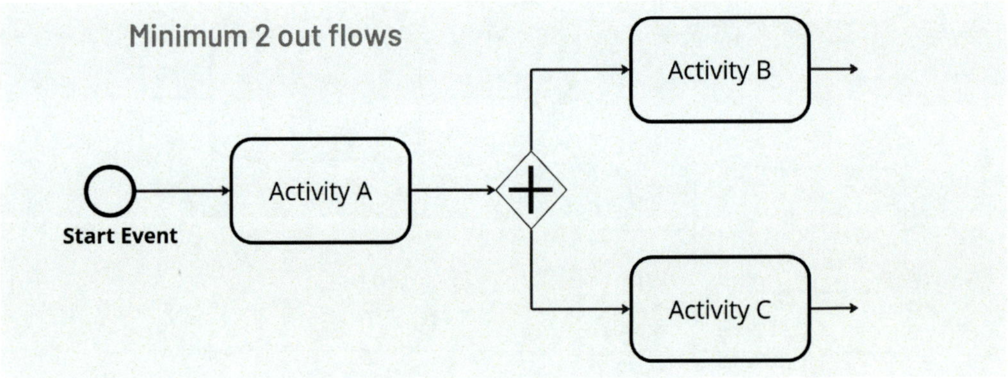

Fg3.6 BPMN Parallel Branching Gateway - Sample with two outflows

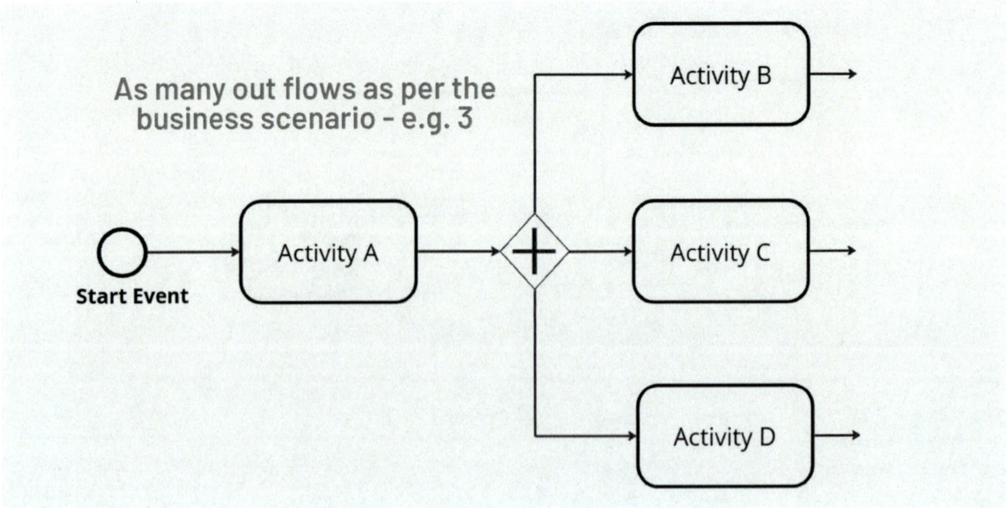

Fg3.7 BPMN Parallel Branching Gateway - Sample with three outflows

Dispelling the Myths of Simultaneous Execution of Parallel Gateway

The Parallel Gateway is often misunderstood as forcing all activities to be executed simultaneously. This is a misconception. In reality, the Parallel Branching Gateway splits the flow into multiple branches that can run in parallel, but they don't have to happen at the same time. The gateway simply allows for multiple tasks to be active at the same time, without imposing any strict concurrency rules.

For example, imagine a hiring process where, after a candidate is shortlisted, two activities are initiated in parallel: conducting background checks and scheduling an interview. While both activities are triggered simultaneously, they don't necessarily happen at the exact same time. The background check might take days, while the interview could be scheduled for the following week. The Parallel Gateway enables these activities to proceed independently but doesn't require them to occur simultaneously. It merely ensures that both tasks must be completed before the process can continue to the next stage, such as making a job offer.

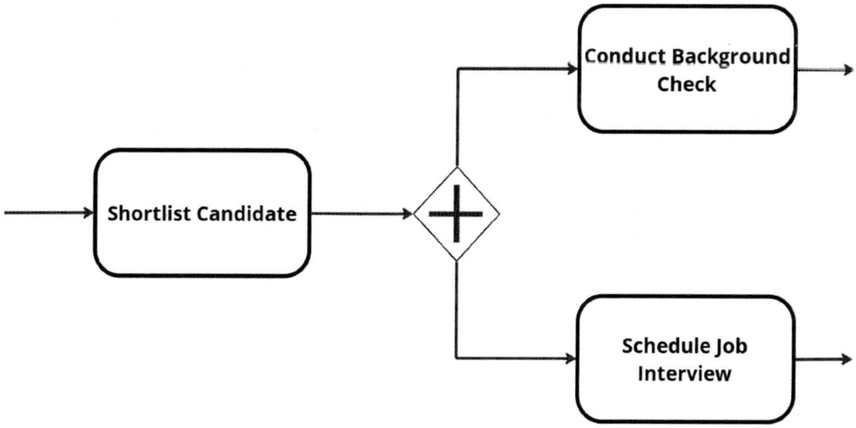

Fg3.8 BPMN Parallel Branching Gateway - Sample with two parallel activities

Parallel Merging Gateway

When used to merge flows, the Parallel Merging Gateway ensures that all incoming paths are completed before proceeding to the next activity. This guarantees that all parallel activities are finished before the process continues, keeping everything aligned and coordinated.

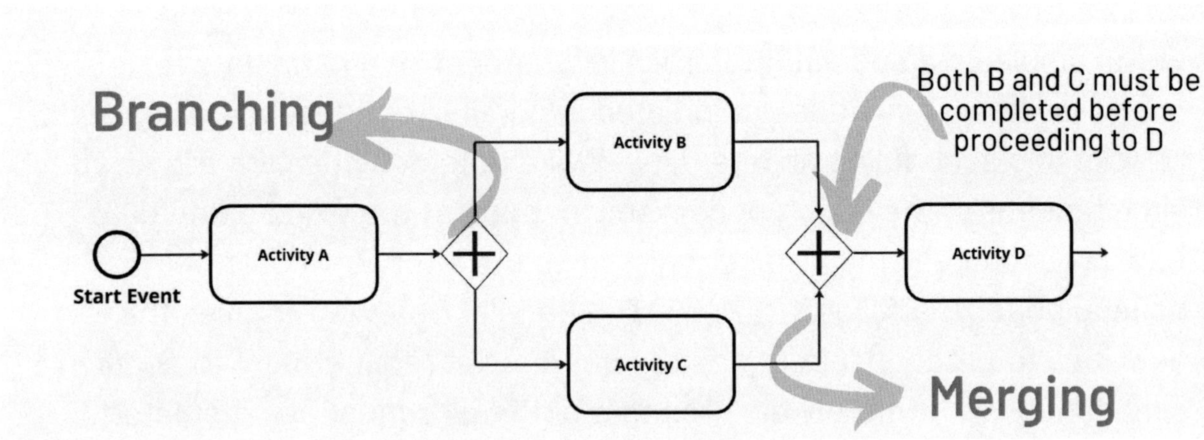

Fg3.9 BPMN Parallel Branching and Merging Gateway - Sample with two parallel activities

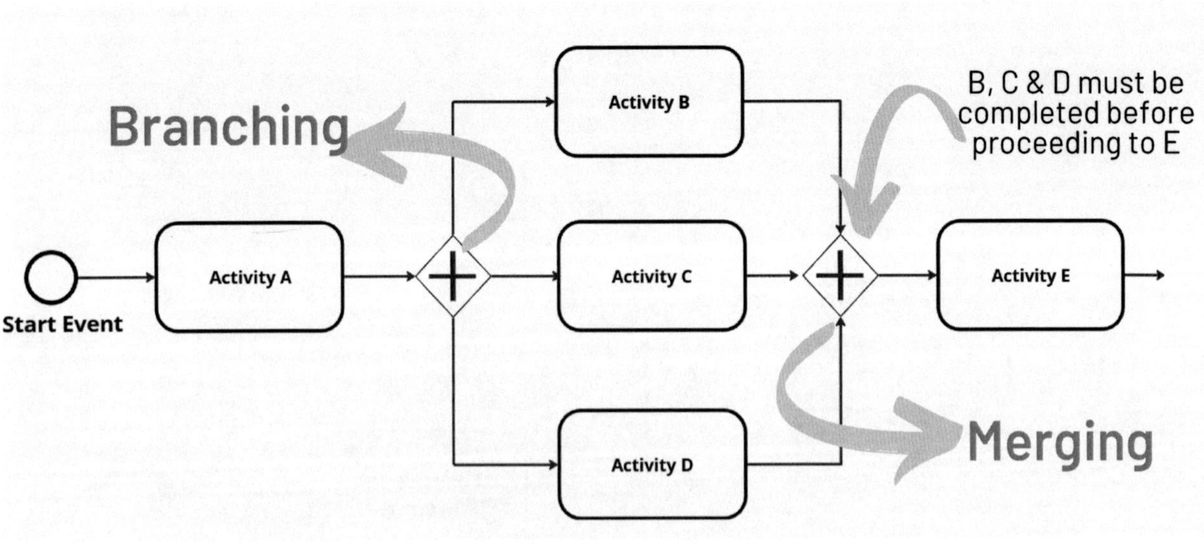

Fg3.10 BPMN Parallel Branching and Merging Gateway - Sample with three parallel activities

In a job application process, the concept of a Parallel Gateway can be used to represent activities that happen concurrently. However, in real life, there isn't an actual "parallel gateway" ensuring everything happens simultaneously or in sync.

For example, consider a job application process where, after an applicant submits their resume, two parallel activities happen:

1. **HR reviews the resume.**

2. **The hiring manager schedules an interview.**

Both activities can occur in parallel, meaning that the HR team and the hiring manager don't need to wait for one activity to finish before starting the other. However, they don't need to happen at the exact same time—HR might take longer to review the resume, while the hiring manager might schedule the interview immediately.

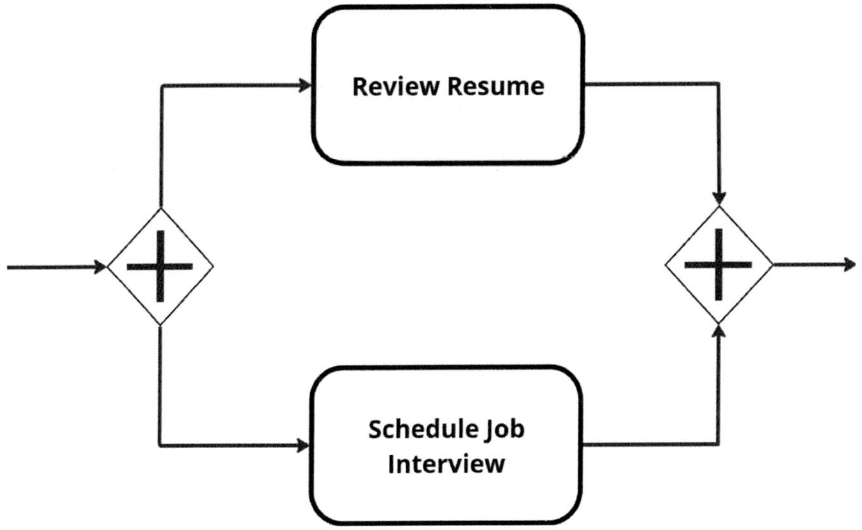

Fg3.11 BPMN Parallel Branching and Merging Gateway - Sample with two parallel activities to illustrate the activities don't need to happen at the exact same time

Merging the Parallel Flows:

Once both activities (Review Resume and Schedule Job Interview) are completed, the process moves to the next step, which could be deciding whether to extend an offer. This is represented by a Parallel Gateway, ensuring that both activities are finished before moving forward.

The Parallel Gateway visually represents this waiting or alignment of activities, but in practice, it's the people or systems that ensure all activities are finished before advancing.

- Human Control: A recruiter or HR coordinator might manually check that both the resume review and interview scheduling are done before advancing to the next step. They act as the "gatekeeper," ensuring all activities are completed.

- System Control: In an automated process, an HR system could track the completion of both activities. The system would wait for confirmation that both the resume review and interview scheduling are complete before allowing the process to move to the next step, such as sending an offer letter or arranging the interview.

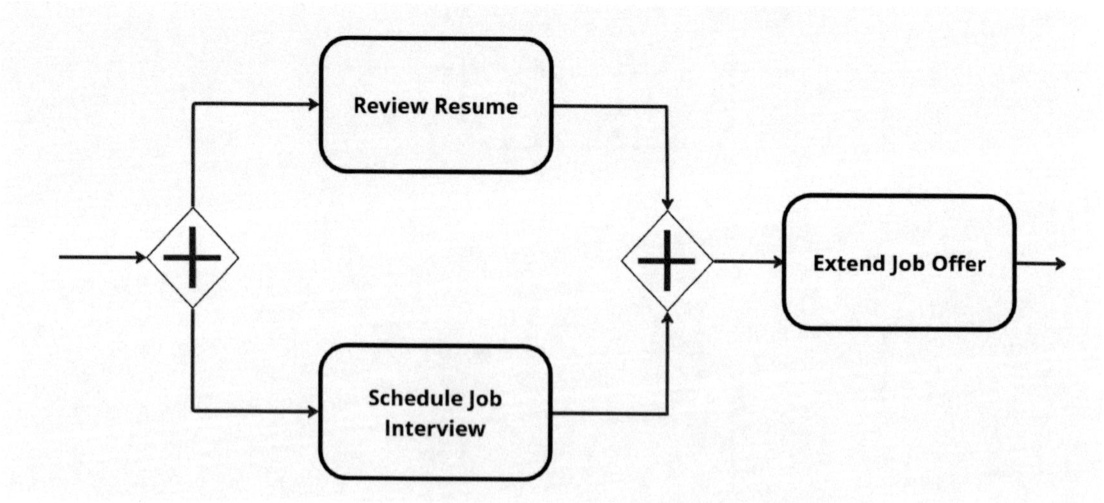

Fg3.12 BPMN Parallel Branching and Merging Gateway - Sample with two parallel activities to illustrate all activities must be first completed before moving to the activity after the Parallel merging gateway

Inclusive Branching Gateway 

An Inclusive Branching Gateway is used to model situations where one or more branches may be activated based on the conditions defined in a process. Unlike the Exclusive Branching Gateway, which allows only one path to be taken, or the Parallel Branching Gateway, which activates all paths, the Inclusive Branching Gateway offers flexibility by allowing any combination of branches to be executed.

An Inclusive Branching Gateway evaluates conditions and then activates one or more paths based on the results of those conditions.

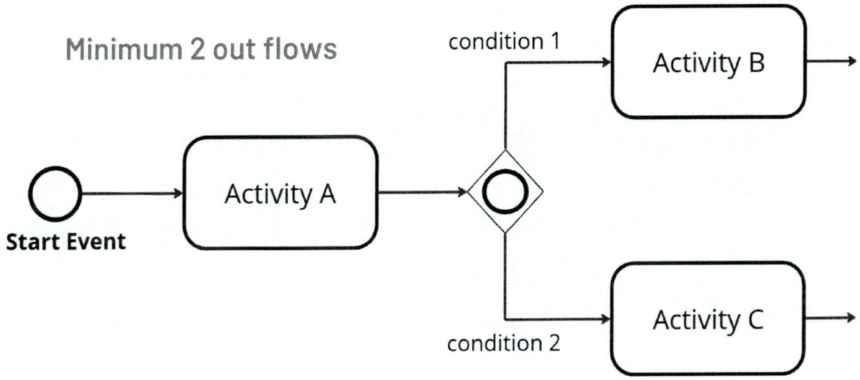

Fg3.13 BPMN Inclusive Branching Gateway - Sample with two outflows

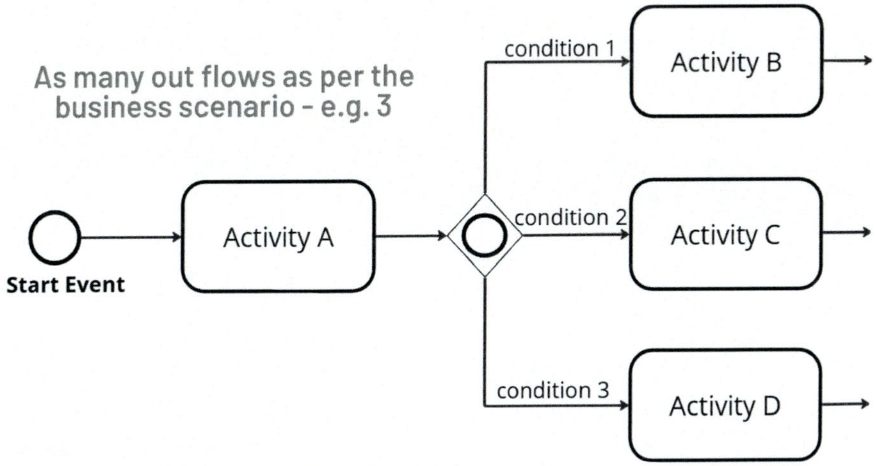

Fg3.14 BPMN Inclusive Branching Gateway - Sample with three outflows

Inclusive Branching Gateway Example in a Job Application Process:

Imagine a job application process where, after an initial interview, several parallel evaluations must occur, but not all are required for every candidate.

1. Technical Test: If the candidate is applying for a technical role, they must take a technical test.
2. Portfolio Review: If the candidate is applying for a creative role, their portfolio must be reviewed.
3. Background Check: All candidates, regardless of the role, must undergo a background check.

Using an Inclusive Branching Gateway allows the process to flow based on specific conditions. For example:

- If the candidate is applying for a technical position, both the technical test and background check will be triggered.
- If the candidate is applying for a creative role, the portfolio review and background check will be activated.

For non-technical, non-creative roles, only the background check is performed.

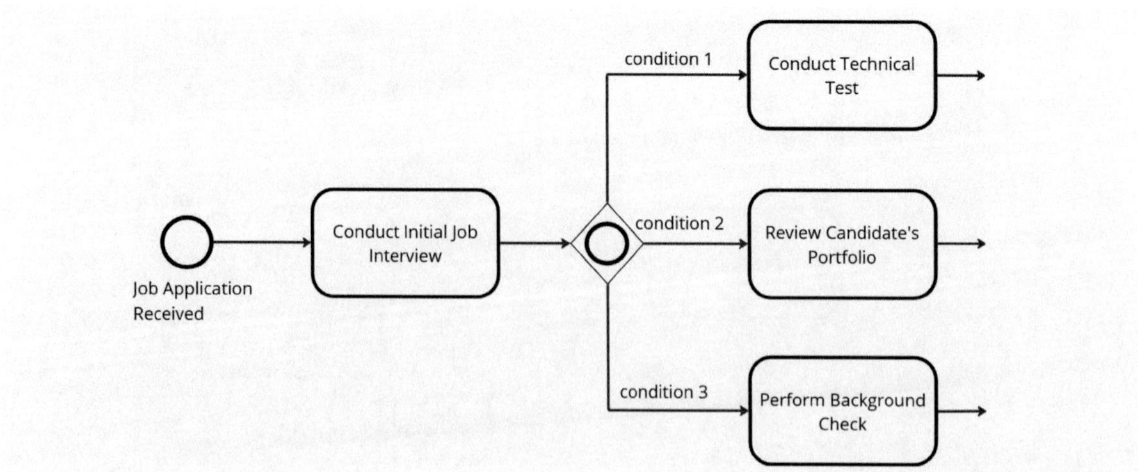

Fg3.15 BPMN Inclusive Branching Gateway - Sample with possible three activities of combinations

Inclusive Merging Gateway:

Once the required activities (e.g., technical test, background check, portfolio review) are completed, the Inclusive Merging Gateway is used to bring the process back into a single flow. This could lead to the final interview step or the decision to extend an offer.

Inclusive Merging Gateway Explained:

The Inclusive Merging Gateway waits for all activated incoming flows to be completed before proceeding to the next task. This is key to its function—it doesn't require all potential incoming flows to be completed, only those that were activated based on the conditions defined earlier in the process. For example, if only the background check and technical test were triggered for a candidate, the process will wait for just those two activities to be completed before moving forward. This ensures flexibility while still maintaining control over the flow of the process.

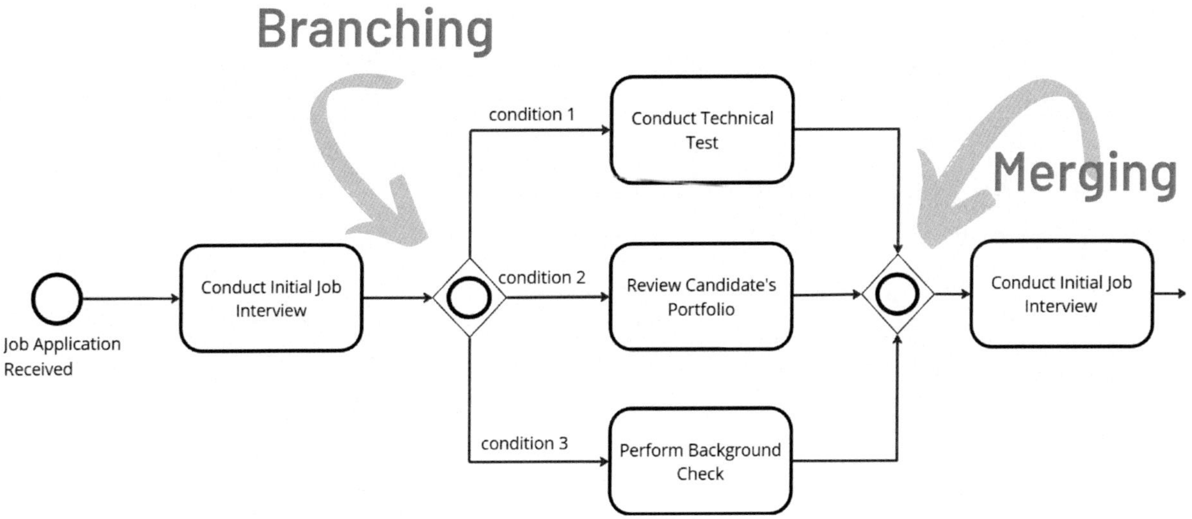

Fg3.16 BPMN Inclusive Branching and Merging Gateway - Sample showing how they work together

Event-Based Gateway 

An Event-Based Gateway is used to model scenarios where the process flow is determined by waiting for one of several possible events. These events can be external or triggered by activities within the process model. Unlike the Exclusive or Inclusive Gateway, where decisions are based on decisions made prior to the Gateway, the Event-Based Gateway waits for the occurrence of one of several events before proceeding along the appropriate path.

The Traffic Light Analogy for Event-Based Gateway:

Imagine you're driving toward a busy intersection with multiple traffic lights. The path you take depends entirely on which light changes first. You don't control the lights, but you react based on what happens:

- If the light turns green, you move forward (triggering one path).
- If a pedestrian crossing signal activates, you wait (triggering another path).
- If the emergency vehicle light flashes, you pull over (triggering a different path).

Just like in Event-Based Gateway, you're simply waiting for one of several events to occur before determining your next action.
The gateway directs the flow of the process based on the event that happens first.

Fg3.17 BPMN Event-based Gateway analogy

Scenario: Job Application Process using Event-Based Gateway for Multiple Response Paths

In this scenario, after the Recruitment Team schedules a job interview, the process waits for different potential outcomes that dictate the next step. Here's how it could unfold using an Event-Based Gateway:

- Candidate confirms the interview (external event).
- No response from the candidate within 5 days (external event).
- HR Director suspends the recruitment process (internal event).

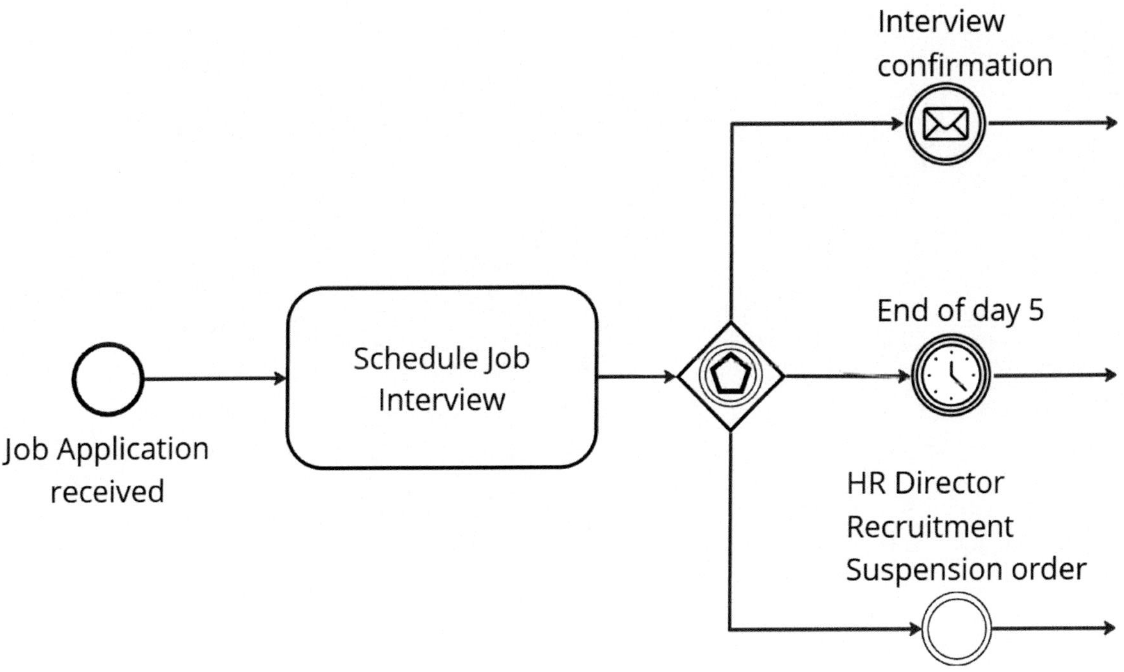

Fg3.18 BPMN Event-based Gateway common usage sample

Each of these events triggers a different path in the process:

- If the candidate confirms, the process proceeds to preparing for the interview.
- If there is no response within 5 days, the process might notify the recruitment team to follow up or take further action.
- If the HR Director suspends the process, all recruitment activities halt, and the team moves to another task or reassesses the job position.

This showcases how the Event-Based Gateway adapts the flow based on various business events, allowing flexibility in handling different real-world outcomes that arise during a recruitment process.

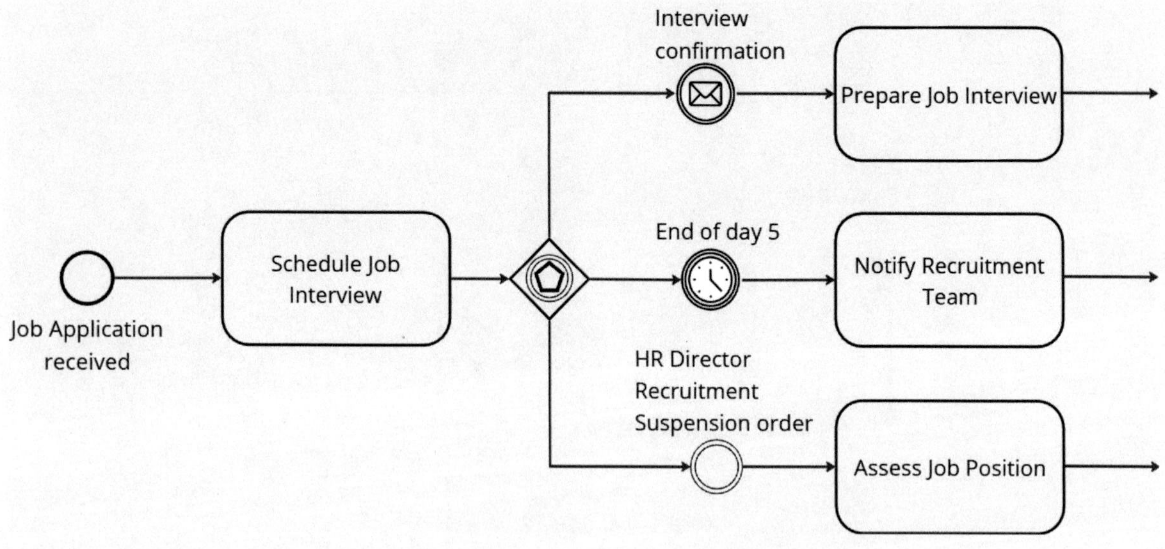

Fg3.19 BPMN Event-based Gateway common usage sample for Job application process

After an Event-Based Gateway, only certain types of Intermediate Catching Events can be used. These events typically represent waiting for a specific signal or occurrence before proceeding in the process.

Here are the possible event types:

1. Message Intermediate Catch Event
   - Incoming message (e.g., receiving a document, a request, or a notification).
   - Example: Candidate confirm the job interview.

2. Timer Intermediate Event
   - Waits for a specific time to elapse or a deadline to be reached.
   - Example: If the candidate doesn't respond within 5 days, this timer event is triggered.

3. Signal Intermediate Catch Event
   - Waits for a broadcast signal that can be triggered by another process or activity within the organization.
   - Example: A signal broadcast when a candidate passes a background check, which triggers the next step in the process.

4. Conditional Intermediate Event
   - Waits for a condition to be true before proceeding.
   - Example: The process waits for the condition that all necessary approvals are completed.

These events represent different "triggers" that can occur and determine which path the process takes after an Event-Based Gateway.

Closing Thoughts

Many people misunderstand Gateways, believing they are responsible for making decisions. In reality, a Gateway is merely a routing mechanism—it does not decide, but rather directs the flow based on conditions that have already been determined before reaching the Gateway.

Think of a Gateway like a traffic intersection. The decision to stop, go, or turn is made by the driver (the process logic), while the traffic lights and road signs (the Gateway) simply enforce the correct route based on that decision. The same applies to BPMN: Gateways do not decide; they simply route the process based on predefined conditions.

Some modelers avoid Gateways altogether, while others overuse them, creating unnecessary complexity. The goal is balance—use Gateways only when they add clarity. Too many Gateways can make a process hard to follow; too few can lead to hidden logic buried inside tasks or flow conditions.

We will explore in Chapter 6&7, on how Process-bility ensures the effective use of Gateways — keeping models clean, intuitive, and free from unnecessary complexity while maintaining clear process logic.

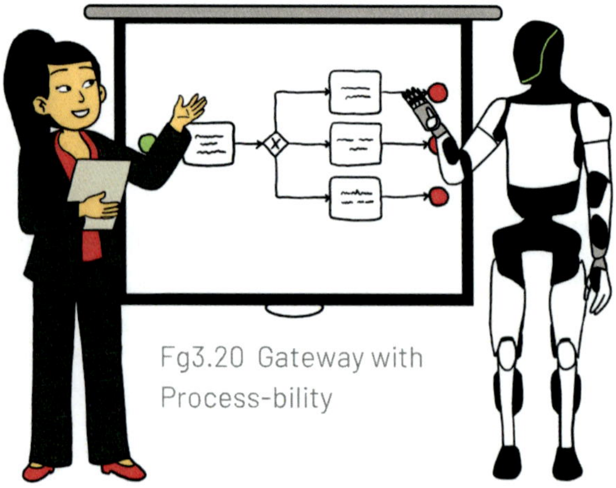

Fg3.20 Gateway with Process-bility

Chapter 4

Visual Balance with Process-bility

- **Visual Balance Transformed My Modeling**
- **Visual Balance Guidelines**
- **Signature Process-bility Visual Balance Element**
    - Person Icon for Swimlanes
    - Process-bility System Representation
- **Why Signature Process-bility Visual Balance Element?**
- **Closing Thoughts**

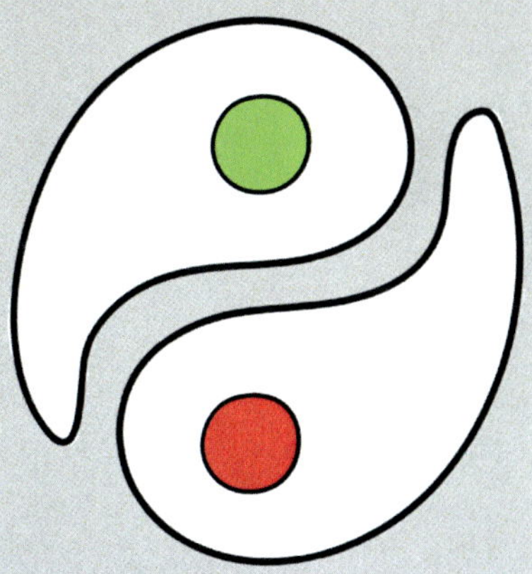

Visual Balance Transformed My Modeling

In 2008, I considered myself a solid business process modeler. My focus was entirely on the content of the model—getting the processes right—but I paid no attention to how they looked visually. Then came a turning point. My project manager, who was also the SME, kept giving me feedback that had nothing to do with the accuracy of the process. Instead, she would point out how the notations were not aligned, or how the connecting flows weren't straight. At first, I found her comments frustrating, as they seemed superficial.

But something changed as I went back to "fix" those visual elements. In aligning the notations and straightening the flows, I realised I had missed a few critical aspects of the process itself. That was my aha moment—the power of Visual Balance.

Visual Balance isn't just about making a model look good; it's about clarity. When a BPMN model is visually balanced, it becomes far easier to spot gaps, inconsistencies, or errors in the process. From that day on, I prioritised Visual Balance in my models, even during process discovery workshops. The impact was immense. Stakeholders could easily follow the flow, provide meaningful feedback, and we could complete the modeling process much faster.

Visual Balance Guidelines

These are the core principles of Visual Balance that ensure your BPMN Models easy to understand:

Left to Right, Top to Bottom Flow:

A left-to-right, top-to-bottom flow is essential. The Start Event should be on the left, with Activities and Gateways aligned naturally from there. This clear direction helps readers follow the process without confusion. When crossing Swimlanes, vertical alignment is key, especially during handovers, but the main rule is to keep the left-to-right flow. Misaligned or randomly placed notations can lead to confusion and make the process harder to follow.

This flow mirrors how we naturally read and process information, making the process narrative easier to grasp. Even complex models become clearer and simpler to understand, allowing the reader to focus on the process itself.

Exception to the Left-to-Right Flow

While left-to-right flow is a general best practice, there is one exception — rework or loop-back scenarios. In cases where an Activity must be revisited due to an error or a business rule (e.g., a review or rework process), it is acceptable for the process flow to move from right to left. This exception makes sense when visually representing a loop back to an earlier point in the process for correction or further action.

Balanced Alignment:

Maintain horizontal alignment (within Swimlane) and vertical alignment (among Swimlanes), ensuring a clean, well-structured diagram that's easy to follow.

Consistent Spacing:

Proper horizontal and vertical spacing between notations keeps the model clean and easy to digest.

Consistent Notation Sizing:

All notations of the same type have identical sizes in your BPMN models.

Consistent Color Coding for Events:

Use green for Start Events, red for End Events, and yellow for Intermediate Events. This uniformity ensures that readers can easily follow the process flow and understand the progression from start to finish.

Avoiding Cross-Overs:

Minimise or eliminate cross-over lines (connecting flows that cross each other). This can often be achieved by rearranging Swimlanes.

Avoiding Excessive Blank Space:

Minimising wasted space in models. Overly large Pool and/or Swim lanes create unnecessary blank areas, which disrupts visual flow and makes the diagram harder to navigate. By keeping pool sizes proportionate, the model remains clean, efficient, and easier to follow.

Avoiding Merging-lines:

Prevents visual clutter, keeping each pathway distinct. This ensures clarity, reduces confusion, and makes the process easier to follow.

Avoiding Unnecessary Zig-Zags in Connecting Lines:

Connecting lines should be smooth and straight to maintain clarity. However, a small exception is allowed when crossing swimlanes to accommodate vertical alignment. In such cases, minimal zig-zagging may be necessary, but the goal remains to keep the lines as direct as possible to ensure a clean and easy-to-follow model.

Person Icons for Role/Team in Swimlanes:

Instead of relying solely on text to represent roles or teams in Swimlanes, use person icons to visually differentiate between roles.

System Representation by Placing the System Label Within the Activity Notation:

Embed the system/app label directly within the activity box to maintain focus on the business process. This ensures clarity and avoids clutter from extra swimlanes or external symbols.

Signature Process-bility Visual Balance Element

These two guidelines go beyond standard BPMN practices and reflect the innovative spirit of Process-bility. They emphasise how process models can be enhanced to improve communication and clarity:

Person Icons for Swimlanes:

Instead of relying solely on text to represent roles or teams in Swimlanes, use person icons to visually differentiate between roles.

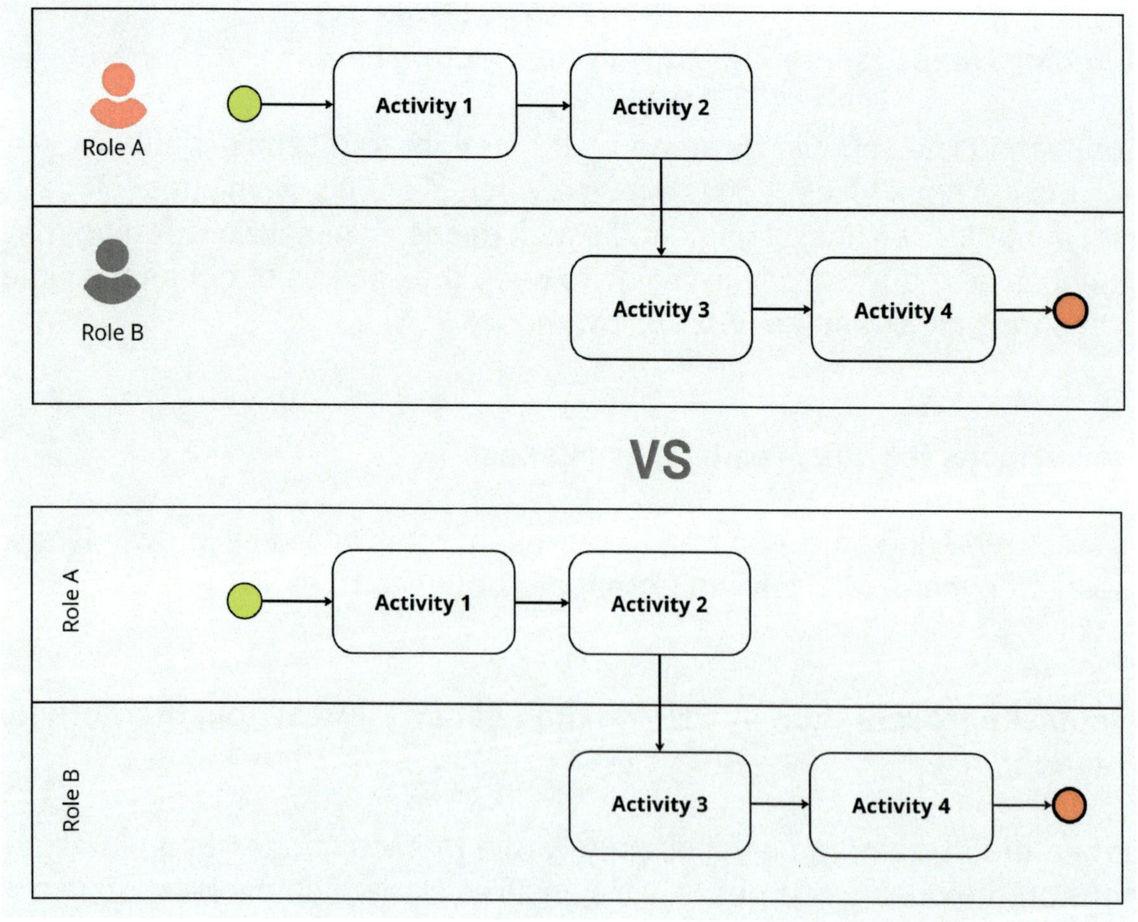

Fg4.1 Showing differences between Process-bility Person Icon and BPMN Role in Swimlane

Why Person Icons Are Superior to Vertical Text?

Immediate Recognition:

Person icons allow stakeholders to instantly recognise the role or team associated with each swimlane. In contrast, the traditional approach of writing role/team names vertically often forces the reader to tilt their head or strain to read sideways, slowing down comprehension.

Better Visual Flow:

The vertical text disrupts the natural left-to-right reading flow, while icons integrate smoothly into the process layout without breaking the visual rhythm. Person icons align with human cognitive patterns, making the diagram more intuitive to follow.

Visual Consistency:

With icons, the roles remain consistent in appearance and placement throughout the model, making it easier for stakeholders to track roles across the diagram. Vertical text, by contrast, varies in length and orientation, often leading to inconsistent visual flow.

Engagement and Appeal:

Icons make the model more visually engaging. They provide a touch of personality and visual diversity, encouraging greater attention from stakeholders during presentations or workshops. Vertical text, especially in complex diagrams, tends to look monotonous and can cause stakeholders to lose interest.

Use of Colors or Variations:

Different colors or variations of the person icon can represent different roles, teams, or departments, improving readability and visual balance. This technique helps stakeholders immediately differentiate between roles, ensuring that the process flow remains clear even in complex models.

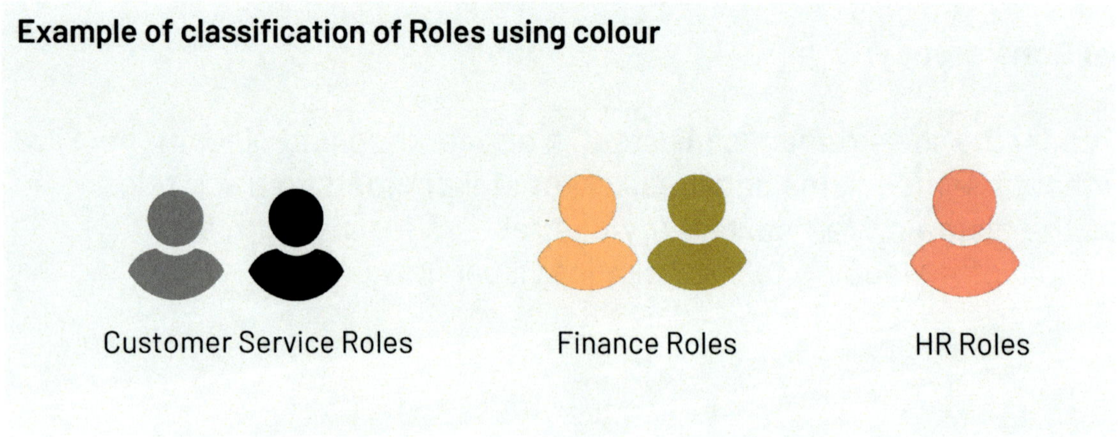

Fg4.2 Sample Person Icon with different colour for logical grouping/association

Process-bility System Representation

Process-bility offers a unique approach to system representation by embedding the system label within the Activity notation. This contrasts with traditional BPMN methods and modern BPM tools, offering significant advantages in terms of clarity, simplicity, and overall Visual Balance.

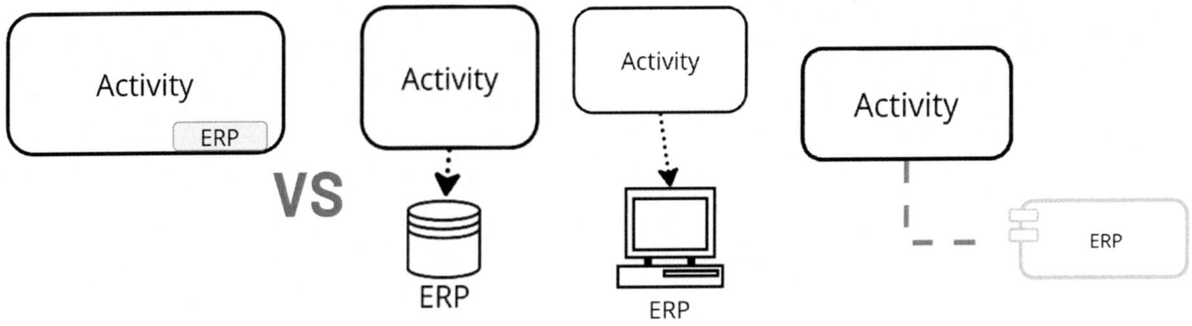

Fg4.3 Process-bility system representation

Advantages of Process-bility Compared to Traditional BPMN Methods:

1. Traditional BPMN Method: Systems as Swimlanes:

- **Challenge:** In traditional BPMN, systems are often represented by creating separate Swimlanes for each system. This makes the process system-focused rather than business-focused, shifting attention away from the core business processes. It also results in more complex and cluttered diagrams, especially when multiple systems are involved.

- **Process-bility Advantage:** By embedding the system label directly within the task, Process-bility keeps the model business-focused. This removes the need for additional swimlanes dedicated to systems, resulting in cleaner, more efficient diagrams where stakeholders can focus on the process itself rather than being overwhelmed by extra lanes.

2. Traditional BPMN Method: Using Data Store Icon:

- **Challenge:** Another traditional method is using a Data Store Artifact (often depicted as a database icon) with association lines to link tasks to the systems they interact with. While this method shows system involvement, it's disconnected from the task, making it harder for stakeholders to immediately see which system is being used.

- **Process-bility Advantage:** In contrast, Process-bility places the system label within the Activity box, providing immediate clarity. Stakeholders don't need to follow association lines or look for external artifacts; the system is clearly visible within the context of the Activity.

3. Modern Method: Linking System with Computer Icon and Association Line:

- **Challenge:** Tools like Signavio often use a computer icon to represent systems, linked to tasks with a dashed association line. While this method visually shows system involvement, it adds complexity by introducing additional icons and lines outside the task. Readers must trace the association line to understand which system is connected to the task, which can be cumbersome, especially in larger diagrams.

- **Process-bility Advantage:** Process-bility eliminates the need for external association lines or icons. By placing the system/app label directly within the activity, it's immediately clear which system is used, reducing visual complexity and improving readability. This direct integration makes it easier for stakeholders to understand the process at a glance without needing to search for linked elements.

Why Signature Process-bility Visual Balance Element?

Clarity and Focus: The Process-bility approach removes unnecessary clutter from Swimlanes, artifacts, or external associations, allowing the system to be immediately visible within the Activity.

Business-Focused Approach: By avoiding the use of extra Swimlanes for systems, Process-bility keeps the focus on the business process rather than making it system-centric.

Simplified Structure: Instead of adding more elements (Swimlanes, icons, or association lines), Process-bility keeps everything contained within the Activity, ensuring that the process flow remains the focal point of the model.

Better Stakeholder Engagement: With systems clearly identified in-task, stakeholders can quickly grasp which tools or systems are involved in each step, allowing for more effective discussions and decision-making.

By embedding system labels directly in the Activity, Process-bility delivers a more intuitive and streamlined approach to system representation, far surpassing both traditional BPMN and modern BPM tool methods in terms of simplicity, business focus, and Visual Balance.

Closing Thoughts

Bringing Visual Balance to Life

We've explored the theory and significance of Visual Balance, seeing how crucial it is for creating clear, effective BPMN models. Visual Balance isn't just a matter of aesthetics; it's a tool that transforms complex processes into easy-to-understand narratives. The real magic, though, happens when these principles are applied to actual models.

In the next chapter, we'll dive deeper into Visual Balance in Action. Here, you'll see real-world BPMN examples and how applying Visual Balance changes the way processes are modeled and understood. We'll take a hands-on approach, demonstrating step-by-step how these techniques can be used to elevate your own models. Get ready for a visual journey that turns complex flows into models that are not only functional but beautiful in their simplicity.

> **Visual Balance is not just about making the models pleasing to our eyes – it unclutters the human mind and helps us crystallise complex processes in seconds.**

Chapter 5

Visual Balance In Action

- **The Easiest Way to Transform BPMN To an Easy to Understand Model**
- **Transforming BPMN Model with Visual Balance**
  - Case 1 : Typical Poor Visual Balance BPMN
  - Case 2 : Cross-over BPMN Model
  - Case 3 : BPMN with Systems
  - Case 4 : Simple BPMN with Re-work
  - Case 5 : All's well that ends well
- **Question and Answers**

The Easiest Way to Transform BPMN To an Easy to Understand Model

Visual Balance is the unsung hero of process modeling. While many modelers focus solely on content, they often overlook how the layout and structure of a model can make or break its clarity. In Process-bility, we believe that every BPMN model should not only convey information but do so in a way that's visually organised and easy to follow.

A well-balanced diagram is like a well-composed piece of art—it draws the viewer's eye naturally, guides understanding, and enhances communication between stakeholders. The best part? Visual Balance is the easiest techniques to apply and delivers immediate results. You don't need to be an expert to see its impact—just a few small changes in alignment, spacing, or flow can dramatically improve your models.

Auto alignment in BPMN modeling Tool

Many BPMN modeling tools, such as MS Visio, Signavio, GBTEC BIC, ARIS, and others, offer auto-alignment or smart/AI alignment features. While these tools can provide some level of assistance in organising models, they often fall short of delivering the true visual balance needed for clarity and understanding.

Automated alignment typically focuses on arranging elements efficiently but may fail to capture the narrative flow, intuitive layout, and structured storytelling that Process-bility emphasises. It's not just about making the model neat; it's about making it make sense at a glance.

Transforming BPMN Model with Visual Balance: Case 1

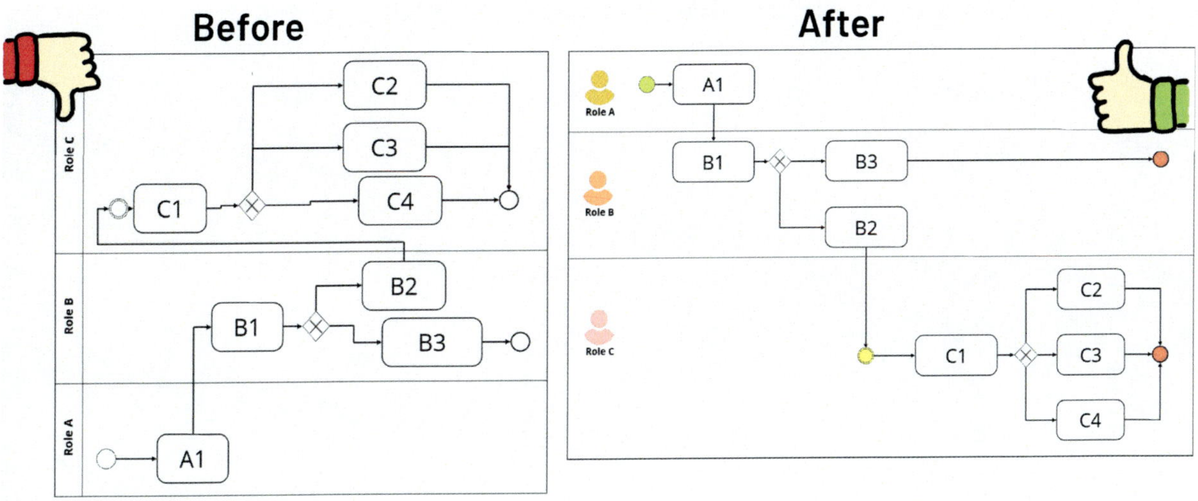

Fg5.1 Visual Balance Case 1

Visual Balance Applied in the After Model:

1. Left-to-Right, Top-to-Bottom Flow: The process now flows naturally from left to right, making it easier to follow and understand.
2. Balanced Alignment: Activities within Swimlanes are horizontally and vertically aligned, providing a clean, structured look.
3. Consistent Spacing: Proper horizontal and vertical spacing between notations makes the model visually clean and easy to digest.
4. Consistent Notation Sizing: All notations of the same type have identical sizes for easy reading
5. Avoiding Excessive Blank Space: Pool and Swimlane sizes have been adjusted to remove unnecessary empty space, ensuring a more efficient layout.
6. Avoiding Merging-Lines: Distinct paths reduce visual clutter and confusion, keeping each flow clear.
7. Avoiding Unnecessary Zig-Zags: Connecting lines are smooth and straight, with minimal zig-zagging, ensuring the flow is easy to follow.
8. Person Icons for Swimlanes: Icons represent each role, enhancing clarity and making it easier to identify responsibilities.
9. Consistent Color Coding for Events: Green for Start Events, yellow for Intermediate Events, and red for End Events help the reader visually track the process progression.

Transforming BPMN Model with Visual Balance: Case 2

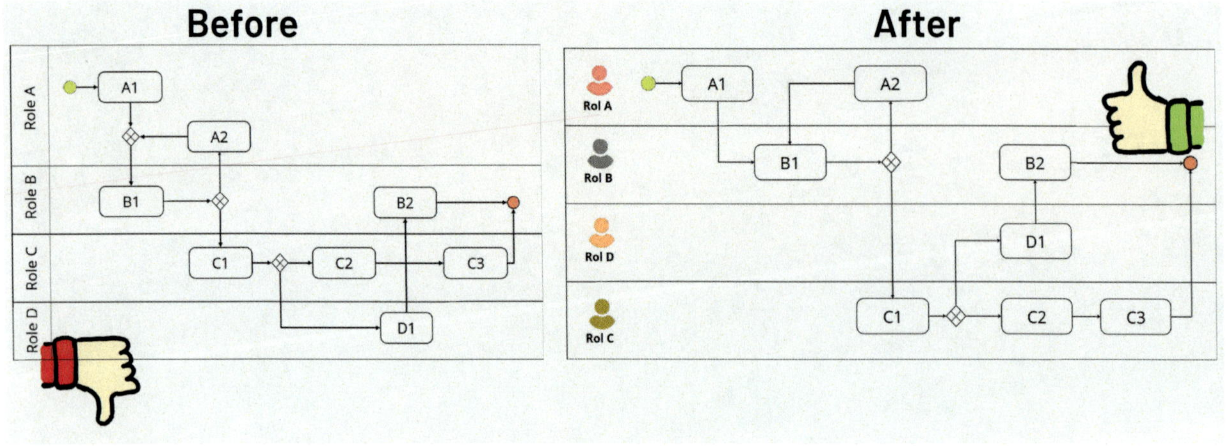

Fg5.2 Visual Balance Case 2

Visual Balance Applied in the After Model:

1. Left-to-Right, Top-to-Bottom Flow: The process now flows naturally from left to right, making it easier to follow and understand.
2. Right-to-Left for re-work/loop back: A2 to B1, removing the Exclusive Merge Gateway
3. Consistent Notation Sizing: C1,C2,C3 and D1
4. Avoiding Cross-Overs: Rearranging Swimlanes (swap Role C with Role D)
5. Person Icons for Swimlanes: Icons represent each role, enhancing clarity and making it easier to identify responsibilities.

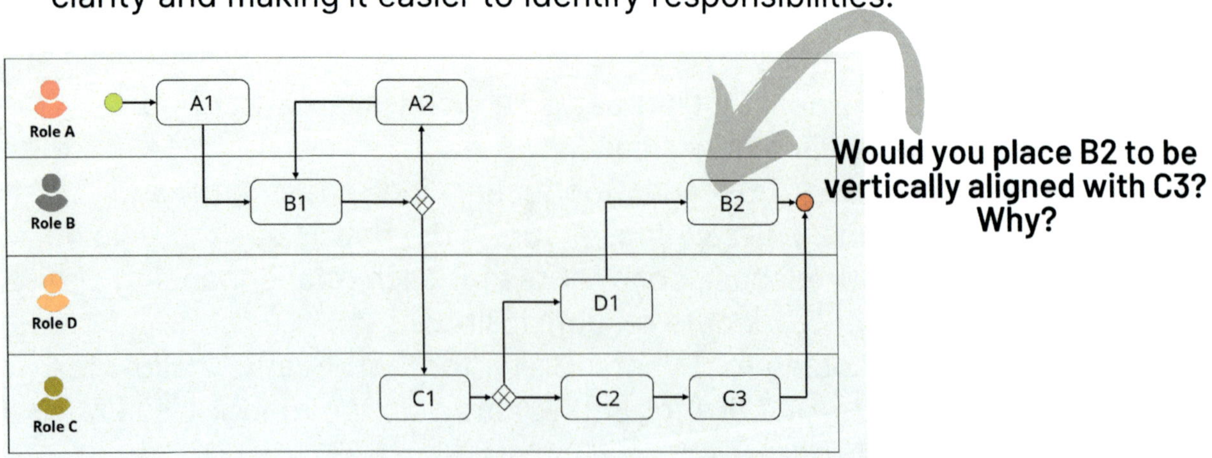

Fg5.3 Visual Balance Case 2 question

Transforming BPMN Model with Visual Balance: Case 3

Before

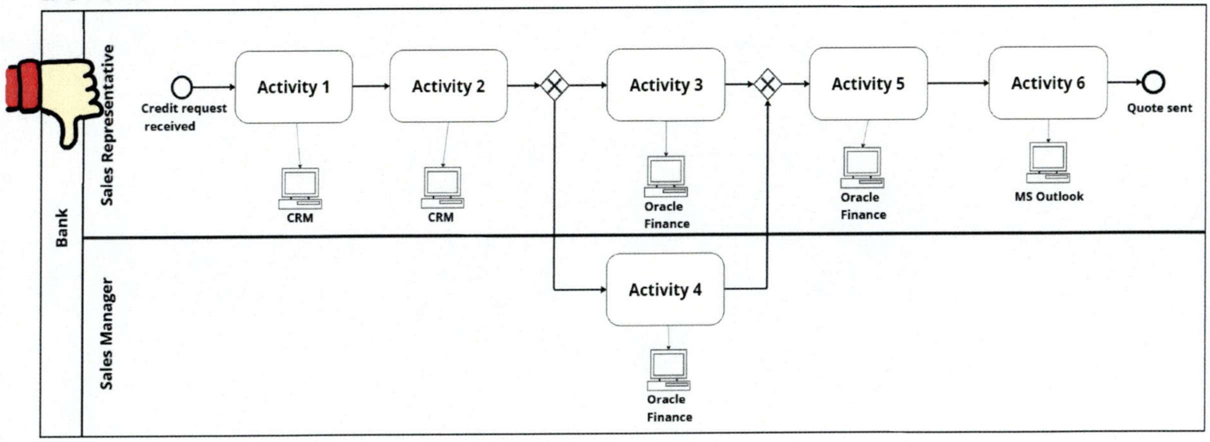

After

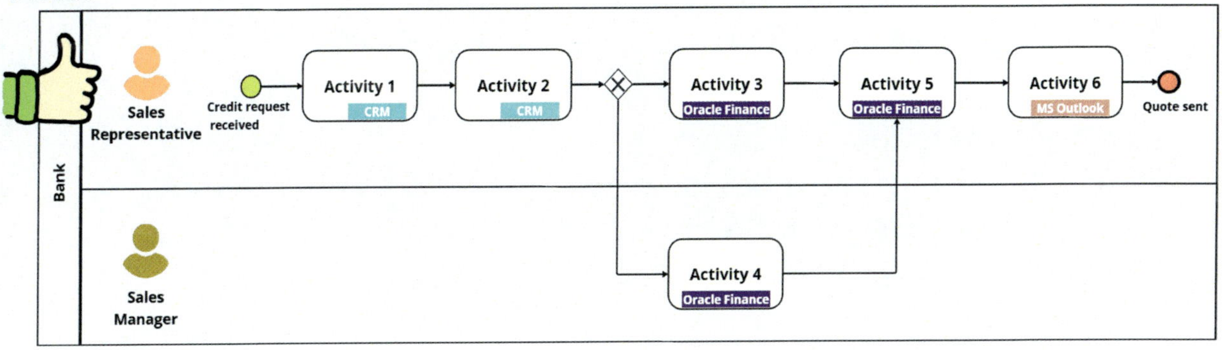

Fg5.4 Visual Balance Case 3

Process-bility Signature Visual Balance Element Applied in the After Model:

1. Person Icons for Swimlanes: Icons represent each role, enhancing clarity and making it easier to identify responsibilities.
2. Process-bility System Representation: Embedding the system label within the Activity notation

Transforming Simple BPMN with Re-work: Case 4

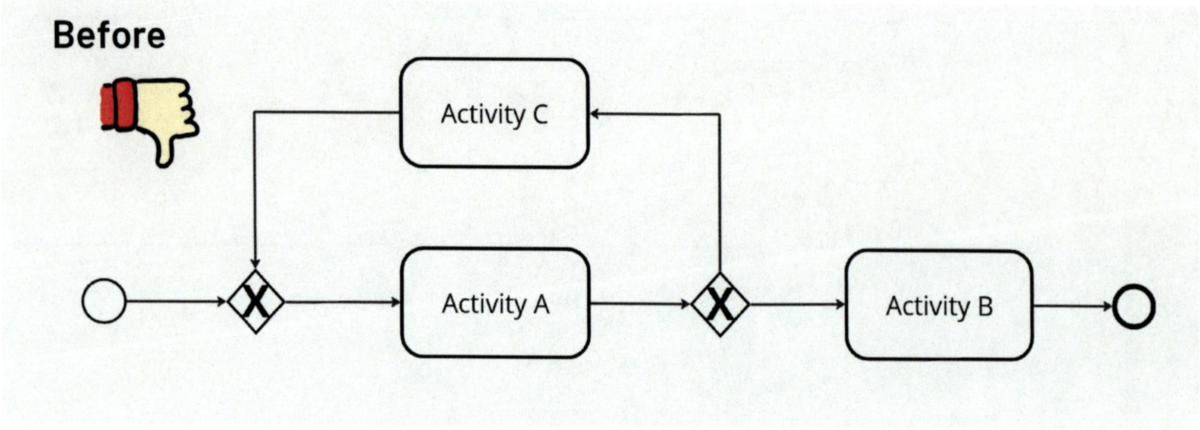

Fg5.5 Visual Balance Case 4 - Before

Hints: Visual Balance Techniques to-be applied:

1. Left-to-Right, Top-to-Bottom Flow
2. Right-to-Left for re-work/loop back
3. Exclusive Merging is not recommended

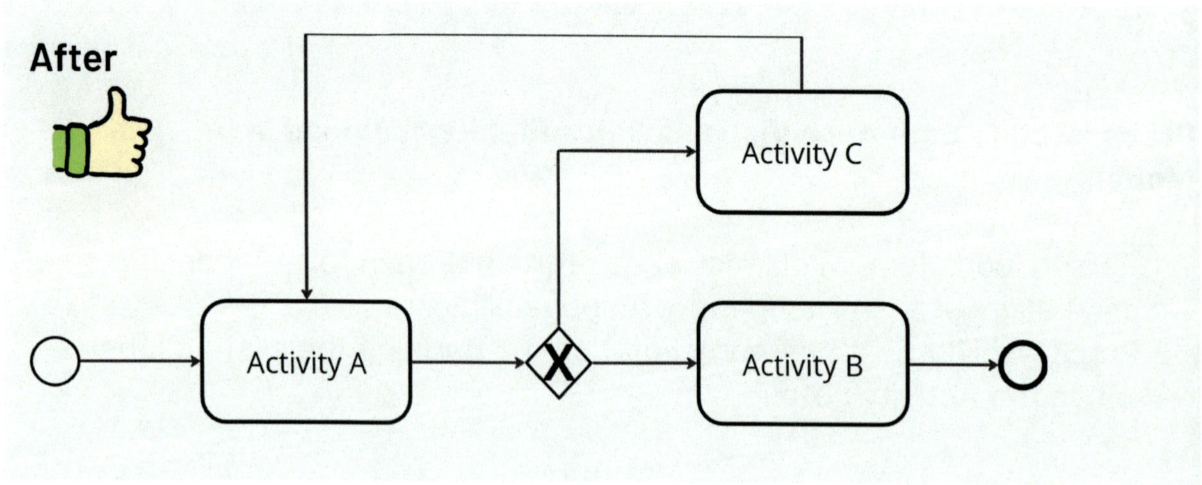

Fg5.6 Visual Balance Case 4 - After

Transforming Simple BPMN Case 5 - All's well that ends well

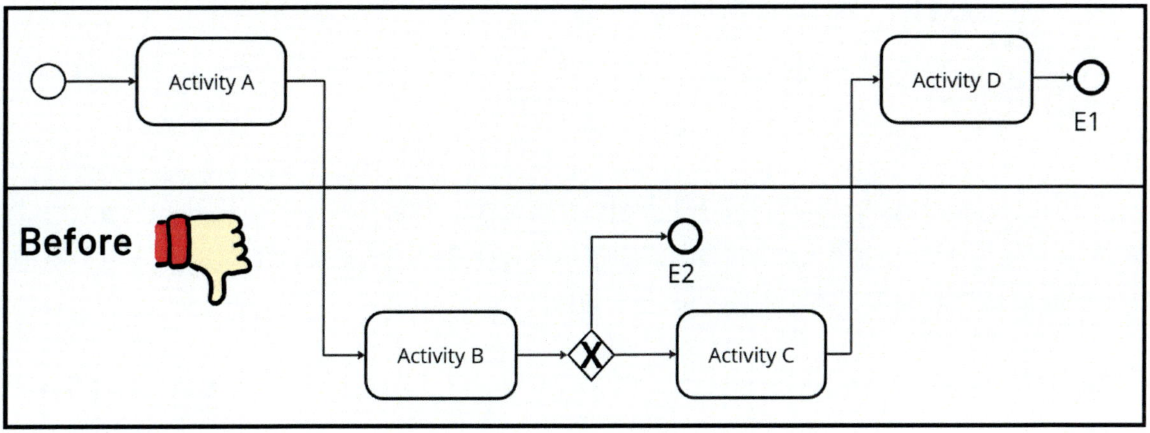

Fg5.7 Visual Balance Case 5 - Before

Hints: Visual Balance Techniques to-be applied:

1. Avoiding Unnecessary Zig-Zags
2. Align multiple End Events vertically whenever possible

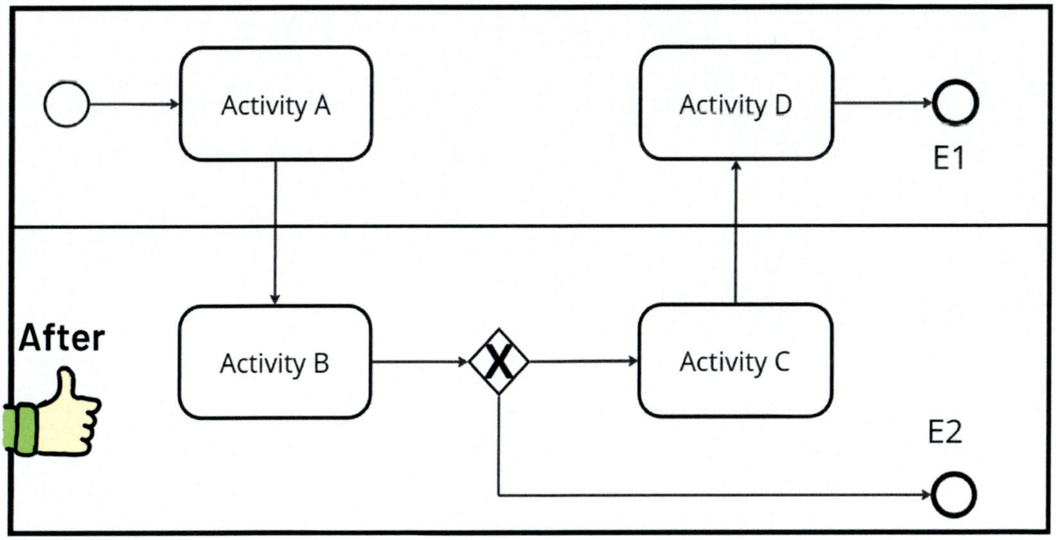

Fg5.8 Visual Balance Case 5- After

When should I apply Visual Balance in BPMN modeling?

Visual Balance should be applied from the very start of your BPMN modeling process. It's an essential part of designing your model, ensuring clarity and simplicity from the beginning. Here's why:

- Clear and Intuitive Flow: By applying Visual Balance early, the process flows naturally, making it easy for stakeholders to follow. This enables them to provide quality feedback sooner and become more engaged in the process workshop, as they can understand the model better.
- Helps the Modeler Understand the Process: Visual Balance isn't just for stakeholders—it also helps the modeler gain a clearer understanding of the business process. By organizing the elements with balance in mind, the modeler can more easily see how each part fits into the overall flow.
- Avoiding Rework: If you wait until the end to fix alignment, spacing, or other balance issues, you risk having to reorganise large portions of your model. Applying Visual Balance consistently saves time and effort.
- Consistency and Simplicity: Visual Balance simplifies even complex processes. By building with balance in mind, you create a structured, clear, and professional-looking model that tells a compelling story from the start.

Visual Balance is a guiding principle throughout your modeling, not just a finishing touch. It improves the quality of your model, enhances your own understanding of the process, and makes the process more engaging and easier to understand for all stakeholders.

Chapter 6

Simplicity In Process-bility

- The Power of Simplicity
- Techniques to Achieve Simplicity
- Closing Thoughts

> **Easy-to-understand BPMN models are the first step toward breaking down the silos in your organization.**

The Power of Simplicity

Simplicity is the key to creating clear, effective process models. By removing unnecessary complexity, we ensure that stakeholders can easily understand and engage with the model. In Process-bility, simplicity is not just about making things look easier—it's about improving communication and decision-making. As Leonardo da Vinci said, "Simplicity is the ultimate sophistication."

Why Simplicity Matters in BPMN Models

Overcomplicated BPMN models lead to confusion, slow down decision-making, and result in wasted time and energy as stakeholders try to decode what should be straightforward. The goal of Simplicity in Process-bility is to reduce cognitive overload, helping people understand processes faster and with greater accuracy.

Models that are simple are more intuitive and invite quicker feedback. This accelerates the entire process of modeling, review, and implementation, making it easier for stakeholders to make informed decisions.

> **Simple but not Simpler.**

Techniques to Achieve Simplicity

a. Streamlined Activity Names

- Action-Oriented Phrasing: The activity names are shortened to focus directly on the action, reducing wordiness.
- Avoiding Unnecessary Details: The names remove unnecessary details like "submitted," "to the candidate," or "for the department," focusing purely on the action itself.
- Direct and Clear: By simplifying the activity names, they are easier to read, process, and understand at a glance, allowing stakeholders to quickly grasp the core actions of the process.

No	Typical Activity Names	Simplicity Version
1	Review the submitted application	Review Application Submitted
2	Complete the background check	Complete Background Check
3	Prepare the budget report for approval	Prepare Budget Report
4	Conduct the interview with the application	Conduct Candidate Interview
5	Assess the quality of the deliverables	Assess Deliverable Quality
6	Approve the budget for the department	Approve Department Budget

Fg6.1 Sample Typical vs Streamline Activity Names

Techniques to Achieve Simplicity (continued)

b. Removing Role Names from the Description of Activity

Roles and responsibilities should already be clear from the Swimlane. There's no need to include the role name in the Activity description. For example, instead of "Manager Approves Invoice," write "Approve Invoice." This avoids redundancy and makes the model cleaner.

No	Typical Activity Names	Simplicity Version
1	Manager Reviews Application	Review Application
2	HR Sends Offer Letter	Send Offer Letter
3	Supervisor Completes Performance Review	Complete Performance Review
4	Team Leader Prepares Budget Report	Prepare Budget Report
5	Coordinator Schedules Interview	Schedule Interview
6	Director Approves Project Plan	Approve Project Plan

Fg6.2 Sample Typical vs Activity Names without Roles

Techniques to Achieve Simplicity (continued)

c. Using Fewer Exclusive Gateways

Wherever possible, simplify your decision points. Rather than using multiple exclusive gateways in sequence, combine them into one and define more specific outflows. For example, instead of two consecutive exclusive gateways, use one gateway with clearly defined paths based on conditions.

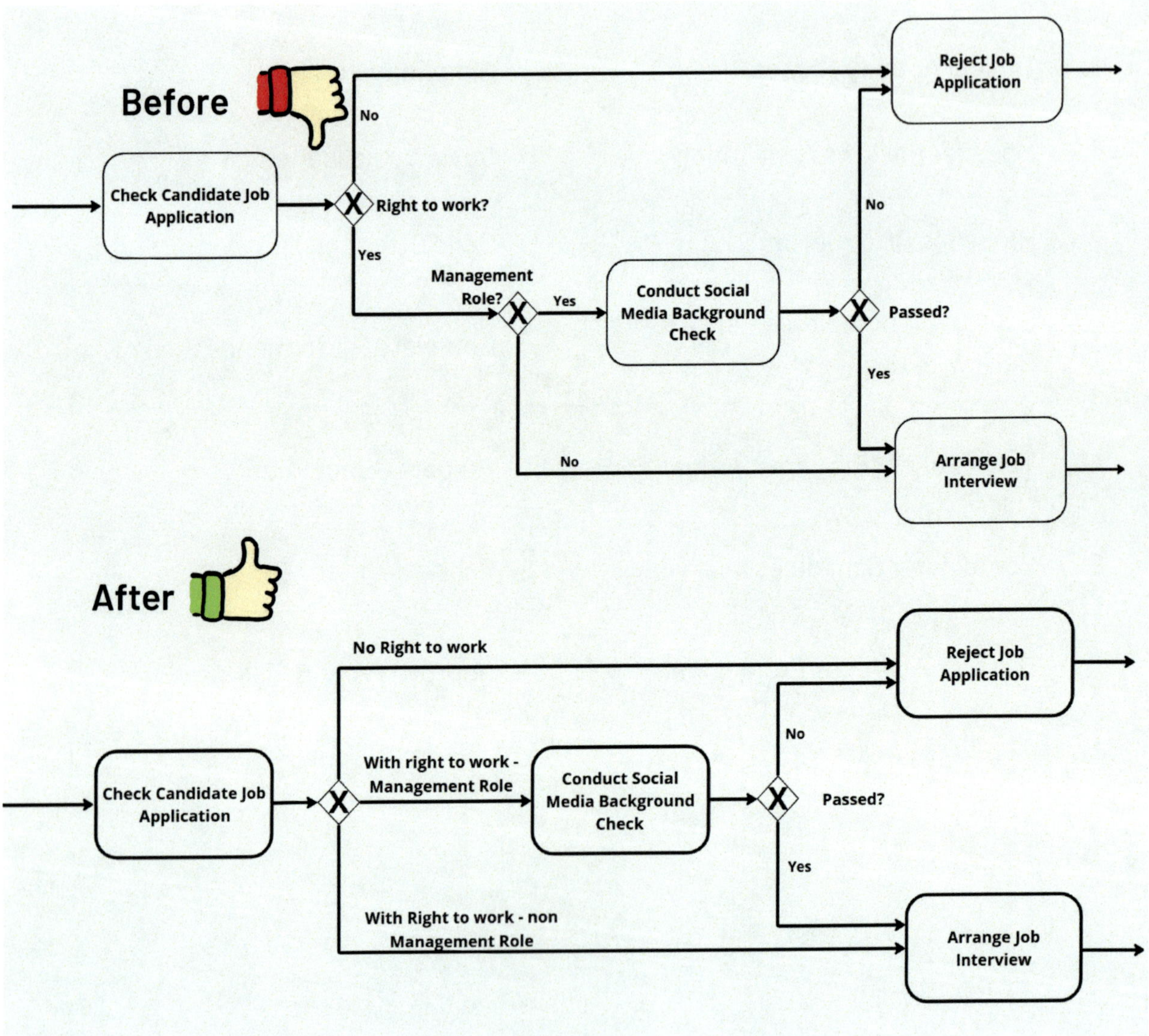

Fg6.3 Reduce Gateways with specific outflows

Techniques to Achieve Simplicity (continued)

d. Using Only Basic Notations

The idea is that less is more. This approach ensures the model stays versatile and clear, focusing on communicating efficiently with fewer notations. In Process-bility, it's not about using more symbols—it's about how clearly you can convey the process. By keeping it simple, stakeholders need less training and can more easily embrace the model, ensuring smoother collaboration and quicker adoption across the organization.

No	Key Elements	Elements
1	Event	UnType Start Event
		UnType End Event
		UnType Intermediate Event
2	Activity	Activity/Task
		Sub-Process
3	Gateway	Exclusive Branch Gateway
		Parallel Gateway
		Inclusive Gateway
		Event-Based Gateway
4	Swimlane	Pool, External Pool, Swimlane
5	Connecting Lines	Connecting Flow, Message and Association

Fg6.4 Basic BPMN Notations

Techniques to Achieve Simplicity (continued)

e. Limiting the Size of the Model to A3 (Monitor-Readable)

A large, sprawling BPMN model is hard to navigate and interpret. Keep your model size limited to an A3 page, which is easily viewable on a standard monitor. This ensures stakeholders can see the full process without constantly zooming in or scrolling.

Fg6.5 Humongous BPMN model that no one bother to understand but the modeler.

Techniques to Achieve Simplicity (continued)

f. Task Types in Process-bility

In Process-bility, the most common task type — User Task — is intentionally left blank instead of showing the person task type icon. This decision emphasises simplicity and avoids visual clutter. Icons for task types should only be displayed when the task is not a User Type. This ensures that attention is drawn to tasks that deviate from the norm, while maintaining simplicity for typical user activities.

g. Don't Mention Systems in the Activity Description

There's no need to mention the system or application in the description of the activity. Systems should be visually represented within the BPMN model itself. This keeps the focus on the action and leaves the system identification to the visual elements in the model, ensuring clarity and simplicity.

No	Typical Activity Names	Simplicity Version
1	Submit Leave Request via HRMS	Submit Leave Request
2	Approve Purchase Order in SAP	Approve Purchase Order
3	Update Customer Information in SalesForce	Update Customer Information
4	Generate Financial Reporting using QuickBook	Generate Financial Reporting
5	Upload Documents to SharePoint	Upload Document

Fg6.6 Sample Activity with system/app

Closing Thoughts

While simplicity is the goal, it's important not to oversimplify. The key is to provide enough detail to represent the process accurately, without overwhelming the model with extraneous information. Process-bility ensures that every detail included is necessary and adds value, striking the perfect balance between clarity and completeness.

Simplicity lies at the heart of Process-bility, ensuring that every BPMN model serves as a clear and actionable guide. By focusing only on essential elements, removing unnecessary complexity, and maintaining visual clarity, we allow stakeholders to engage with the process swiftly and effectively. Simplicity doesn't just make models easier to read — it empowers organizations to act with precision, enabling continuous improvement and faster problem-solving.

Simplicity is not just a principle — it's a practice. In the next chapter, Simplicity in Action, we'll dive into real-world examples and explore before-and-after models that demonstrate how these techniques transform complex diagrams into intuitive, streamlined workflows. Let's see how applying simplicity in BPMN modeling can unlock the true potential of your processes.

Chapter 7

Simplicity In Action

- **Transforming BPMN Model with Simplicity**
    - Case 1 : Typical Complicated BPMN Model
    - Case 2 : User Task Type with lengthy description
    - Case 3 : A Super huge spaghetti BPMN Model
    - Case 4 : Common Approve/Reject bad practices
    - Case 5 : Get Away with Gateway

Transforming BPMN Model with Simplicity: Case 1 - Before

Before 

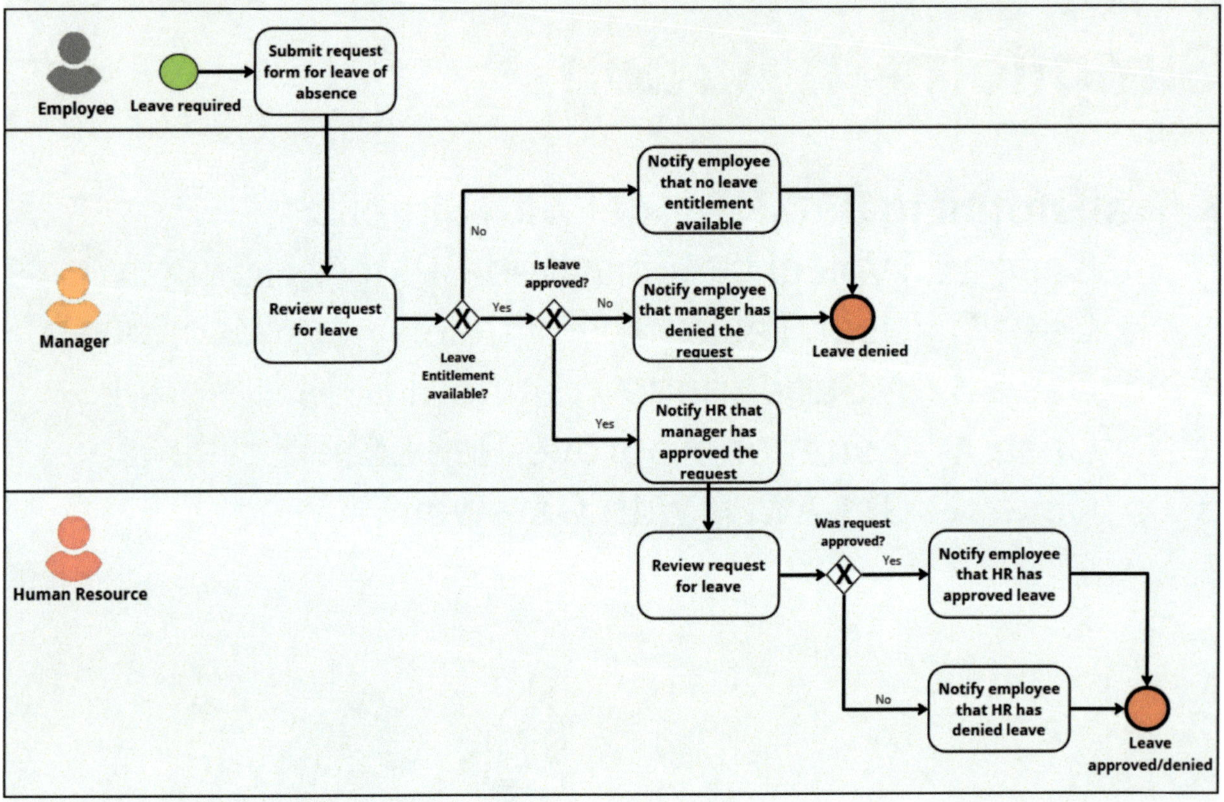

Fg7.1 A Before Leave Application Process

This "Before" BPMN model illustrates a leave request process that includes multiple decision points and notifications. While the model is technically correct, it contains redundant elements, excessive gateways, and overly detailed steps that complicate the flow. This level of complexity can make it difficult for stakeholders to follow the process at a glance. By applying Simplicity techniques, we can refine this model to focus on the essential tasks, streamline the decision paths, and eliminate unnecessary elements, resulting in a clearer, more efficient representation of the process.

Transforming BPMN Model with Simplicity: Case 1 - After

After

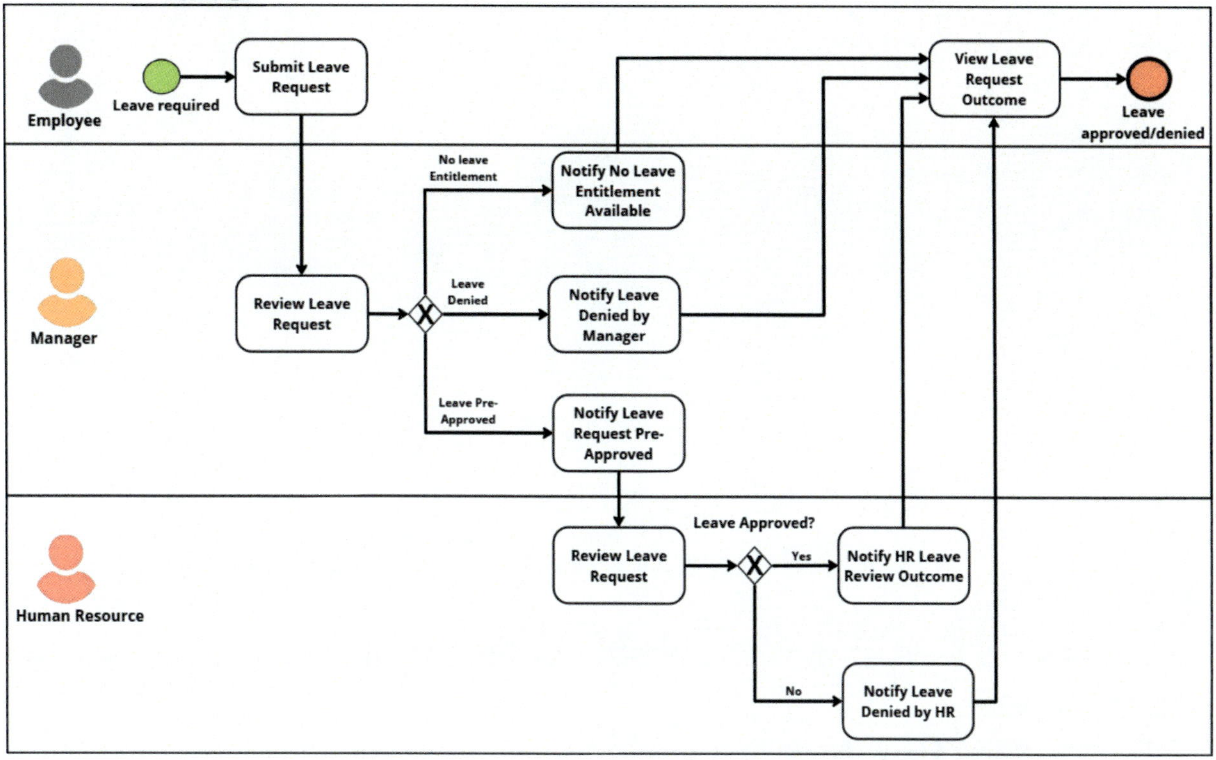

Fg7.2 An After Leave Application Process

In the "After" model, key Simplicity techniques applied include:

1. Streamlined Activity Names - Reducing unnecessary filler words for clearer, concise labels.
2. Removing Role Names from Activity Descriptions - Since roles are visually assigned in Swimlanes, role names within activities are redundant.
3. Using Fewer Exclusive Gateways - Reducing the number of gateways to create a more streamlined flow, making the model easier to follow.

These changes result in a cleaner, more intuitive model that is quicker to understand and engage with, allowing stakeholders to focus on essential decision points and outcomes. This approach is particularly useful in complex workflows where clarity is crucial for both communication and effective process management.

Transforming BPMN Model with Simplicity: Case 2

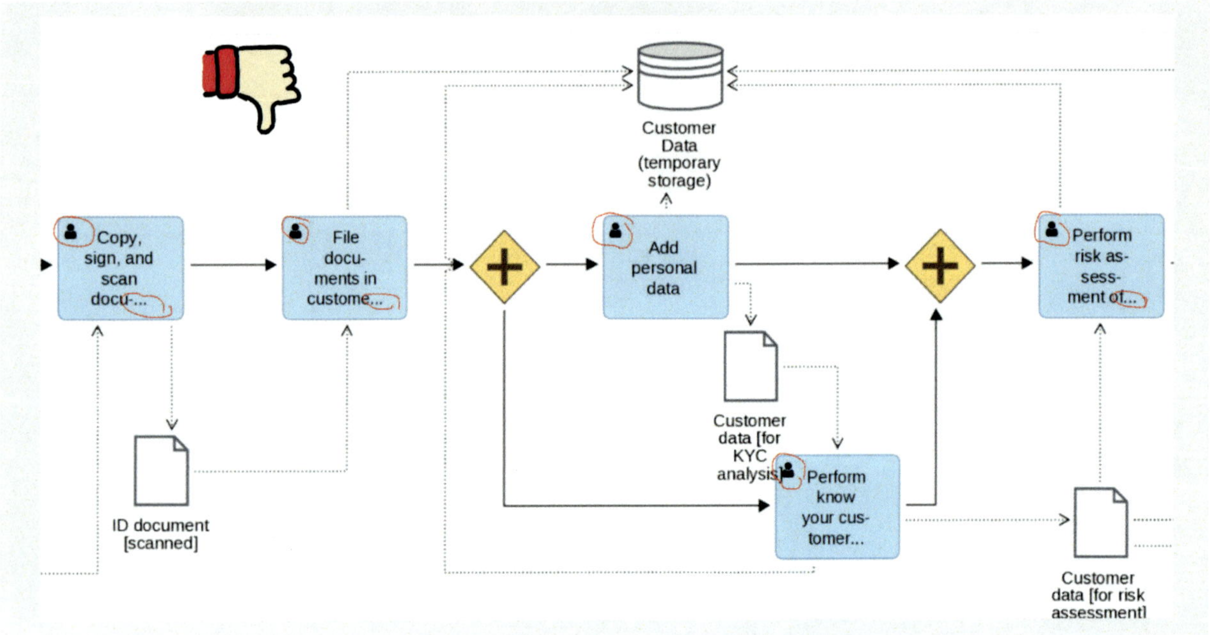

Fg7.3 Actual model with Unnecessary User Task Type and long activity description

This BPMN model illustrates an opportunity for simplification by addressing two key areas:

1. **Omitting the User Task Type:** Since User Tasks are the most common task type in this model, they can be omitted from the activity icons to reduce visual clutter. This allows viewers to focus on the process flow without being distracted by repetitive symbols.
2. **Streamlining Task Name:** The Task Names currently exceed the space within the Task boxes, making the model harder to read. By using shorter, more concise Task Names, we can create a cleaner, more organised layout. For example, "Copy, sign, and scan documents" could simply become "Archive ID Document." This approach improves readability, allowing stakeholders to quickly grasp each step of the process.

Applying these techniques will create a model that is both visually streamlined and easy to understand, aligning with Process-bility's emphasis on simplicity and clarity.

Transforming BPMN Model with Simplicity: Case 3

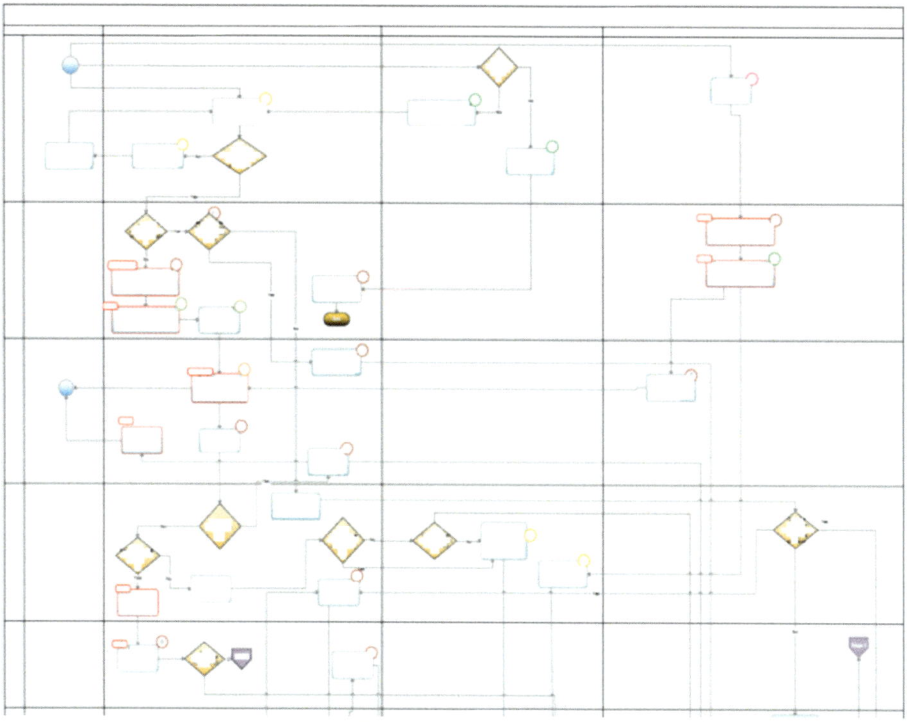

Fg7.4 An Actual Complaints Management Process of a finance institute (Published with Permission - Details Masked)

This Complaints Management process model is an example of how excessive complexity can render a BPMN model ineffective. With numerous gateways, overlapping paths, and scattered notation, the visual clutter makes it challenging for any reader to follow or understand the process flow. Even though it technically represents the complaint-handling steps, the overuse of notations and detailed paths creates confusion rather than clarity.

When BPMN models reach this level of intricacy, they fail to serve their purpose as a communication tool. Instead, they become a barrier to comprehension, limiting the model's utility for stakeholders who need actionable insights. Through Process-bility, we aim to transform such models into streamlined, reader-friendly visuals by focusing on simplicity, storytelling, and clarity — ultimately turning complex processes into actionable, intuitive insights.

Transforming BPMN Model with Simplicity: Case 4

Imagine you are a Business Analyst tasked with modeling the business process for a job application review. In this process, the Hiring Manager is responsible for reviewing each candidate's application. After the review, the manager must make a decision on whether to approve or reject the application.

- Approval: If the application is approved, the Hiring Manager schedules the candidate for an interview.
- Rejection: If the application is rejected, the Hiring Manager notifies the candidate of the unsuccessful outcome.

Take a moment to create your model based on this scenario. On the next page, we'll explore a typical approach and then look at the Process-bility approach to see how Simplicity can improve clarity.

Transforming BPMN Model with Simplicity: Case 4

Typical BPMN Model

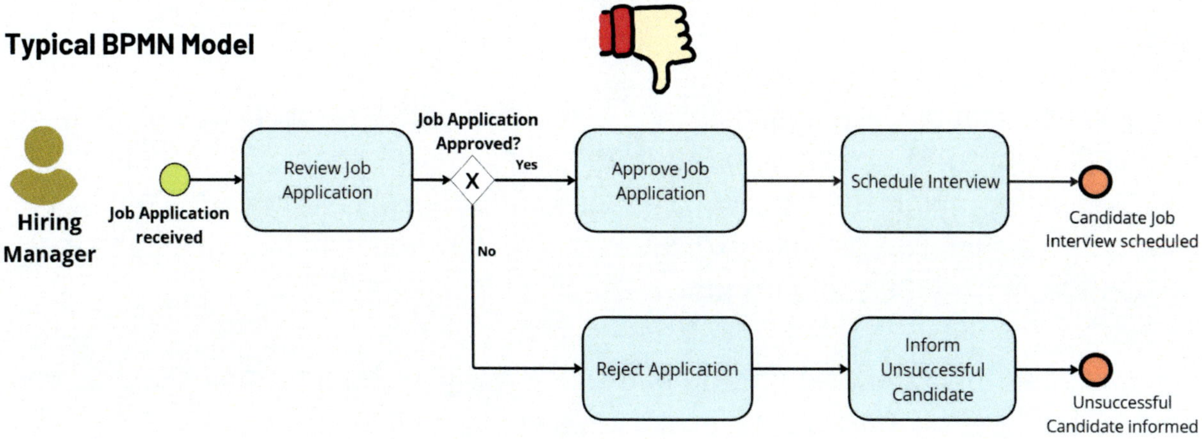

Fg7.5 Typical overly detail BPMN Model - approve/reject

We see separate Activity for "Approve Job Application" and "Reject Application" after the initial "Review Job Application". However, these decisions are actually part of the review process itself, making the extra steps redundant. This leads to an overcomplicated model with unnecessary tasks.

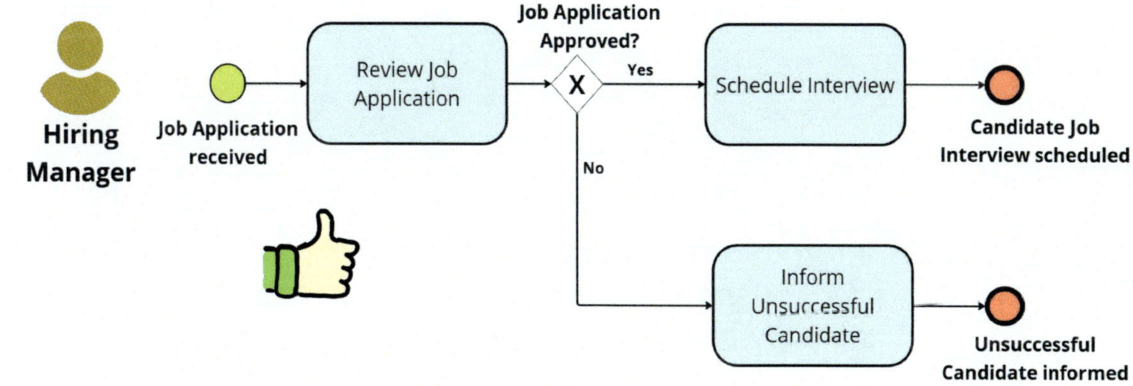

Fg7.6 Process-bility approach - Approve/Reject is part of the Review Activity

With Process-bility, the Simplicity principle is applied by merging the approval and rejection actions directly into the "Review Job Application" task. Now, the decision gateway directly leads to either scheduling the interview (if approved) or informing the unsuccessful candidate (if rejected). This streamlined approach reduces the number of tasks and clarifies the process flow, making the model easier to read and understand without losing any essential information.

Transforming BPMN Model with Simplicity: Case 5 - Get away with Gateway

In a job application process, after verifying a candidate's qualifications, a Police Background Check is mandatory for every applicant. Additionally, candidates applying for certain roles may require a Working with Children Check. If all required checks are passed, the candidate moves forward to the Job Interview Scheduling stage.

What makes this scenario unique is the flexibility in the sequence of checks. The Police Background Check and the Working with Children Check can occur before, after, or even simultaneously, depending on workflow demands and resource availability. If a candidate fails either the Police Background Check or the Working with Children Check, they will receive a notification of rejection. This adaptable setup challenges traditional process modeling approaches, often resulting in overly complex BPMN models that rely on multiple exclusive gateways and fail to capture the flow's natural flexibility.

Challenge for You:

Before moving on, take a moment to model this scenario yourself. Consider the nuances—the flexibility in sequence, the mandatory checks, and the potential outcomes. Try to capture this in BPMN without looking at the next page. Remember, your model should communicate the process clearly and concisely, with a focus on both Simplicity, Visual Balance and accuracy.

Give it a try! Once you're ready, turn the page to see a typical BPMN approach and then the streamlined Process-bility version.

> **Today, the winners are those who have the best process, but tomorrow belongs to those who understand their process.**

Transforming BPMN Model with Simplicity: Case 5 - Get away with Gateway

Typical BPMN model

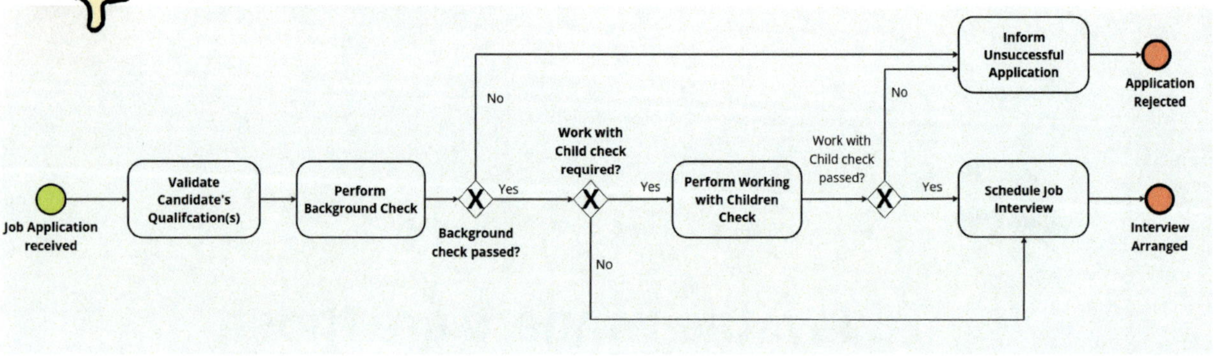

Fg7.7 Over-simplify the business process with complicated BPMN model

BPMN model with Process-bility

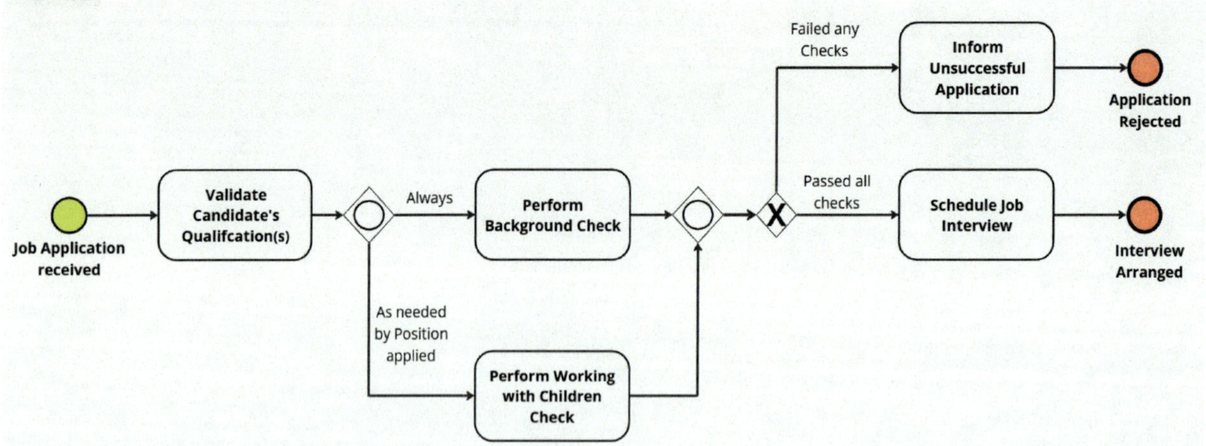

Fg7.8 Model complex business process with simple model - simple but not simpler

Chapter 8

Storytelling Through Process-bility

- **Storytelling Bring Process Models to Life**

- **Process-Structure Storytelling Techniques**
  - Identifying the Object of Focus
  - Crafting the Journey
  - Analogy: The Lord of the Rings
  - Decomposition With Sub-Processes

- **Modeling-Specific Storytelling Techniques**
  - Verb + Object Naming for Activities
  - Noun + Past Participle for Events
  - Avoid Role Name in Activity Naming
  - Meaningful Start and End Event Naming
  - Use of Intermediate Un-type Events
  - Aligning Start and End Events Across Linked Process Models

- **The Benefits of Storytelling in BPMN**

- **Closing Thoughts**

Storytelling Bring Process Models to Life

At the heart of every effective BPMN model is a business story told with clarity. Storytelling is the foundation of Process-bility, and it's how we make sense of processes, align stakeholders, and drive real change. However, storytelling in BPMN isn't just about stringing together tasks and activities—it's about conveying the journey of an object of focus, ensuring clarity, and making processes understandable for everyone, from business analysts to business users.

Fg8.1 Storytelling bring process models to life

In Process-bility, storytelling involves two key aspects: Process-Structure Techniques and Modeling-Specific Techniques. The combination of these two approaches ensures that your process models are both functional and engaging. By structuring your process clearly and using consistent naming conventions, visual balance, and simplicity, you bring the story of the business process to life.

Process-Structure Storytelling Technique

When creating BPMN models using Process-bility, the structure of your model plays a crucial role in telling the story of the process. The way processes are broken down, linked, and visualised provides the audience with a clear path to understanding the sequence of activities, roles, and outcomes. These structural choices help create a process narrative that stakeholders at all levels can follow, ensuring clarity from start to finish.

At the core of this structure is the object of focus — every process has one. By centering your model around this key object, you maintain a clear, engaging narrative.

The following Process-Structure techniques ensure your models don't just document tasks—they tell a coherent story that guides stakeholders through each stage of the process.

1. Identifying the Object of Focus

Every story has a main character, and in the context of BPMN, the main character is the object or entity being tracked or processed. This could be anything from an invoice, a customer order, a contract, or even a complaint.

In each process or sub-process, there is always at least one object that is the focal point. Identifying this object helps to structure the narrative and keeps the process clear. For example, in an invoice approval process, the invoice is the central object. From the start of the process (invoice submission) to the end (payment or rejection), every activity influences the state of this object.

When crafting a BPMN model, it's crucial to identify the main object of focus to ensure that the process revolves around it. This approach provides clarity to the reader and ensures the process remains aligned with its purpose.

Fg8.2 Object of focus

2. Crafting the Journey

Once you've identified the object of focus, the next step is to craft the journey it takes through the process. This is where storytelling and Process-bility truly shine.

A process, like a story, has a beginning, middle, and end. The goal is to map out the journey the object takes from one state to another. To craft this journey effectively, it's important to ensure the process is clear, logical, and structured in a way that makes sense.

Here are some principles for structuring processes for maximum clarity:

- Start with the beginning: Where does the process start? What event triggers the process?
- Clarify the flow: What activities or tasks does the object go through as it moves forward?
- Identify key decisions: Where are the decision points, and how do they impact the flow of the process?
- Keep the end in mind: What is the final outcome? Does the process resolve, and if so, how?

By treating the process like a narrative, we ensure that it is both understandable and effective for its intended audience.

Fg8.3 Journey To The West

3. The Lord of the Rings Analogy

To better understand how storytelling object of focus works in BPMN, let's draw on a familiar analogy: The Lord of the Rings.

Fg8.4 LOTR Analogy

In this epic story, the Ring is the central object of focus. The journey revolves around taking the Ring from the Shire to Mount Doom, where it will be destroyed. Along the way, different characters interact with the Ring: Frodo carries it, Gandalf advises on it, and the villains are chasing after it. Despite all the characters and events in the story, the journey of the Ring remains the focal point.

Similarly, in BPMN, the "Ring" of your process is the object you're tracking, like a customer order or invoice. The journey the object takes through the process is crucial. Along the way, various roles and tasks influence the state of the object, but the object itself remains central.

Just as the journey of the Ring guides the narrative of The Lord of the Rings, the journey of the object guides the narrative of your BPMN model. Understanding this helps create clarity in your process models and makes the story easier to follow.

4. Decomposition With Sub-Processes

The essence of business process modelling is the art of composing activities at the right level of details. One of the most powerful techniques in Process-bility is the use of decomposition — breaking down complex processes into smaller, more digestible parts through the use of sub-processes.

Think of decomposition like breaking a book into chapters. Each chapter (or sub-process) contains a specific part of the story, but all contribute to the larger narrative. This is especially useful when dealing with complex workflows or processes that span multiple departments or require detailed tasks.

When to Use Sub-Processes

- The process is too large to fit clearly on a single page/screen (About A3 in size).
- Different teams or departments handle distinct sections of the process.
- You need to drill down into more detailed tasks without cluttering the main process model.

How to Use Sub-Processes

When using sub-processes, the trick is to ensure that activities on the same "level" of detail are grouped together. Think of this as ensuring that each chapter in a book tells a cohesive part of the story.

By keeping the details of each sub-process within its own "chapter," you prevent the main process from becoming too cluttered while still providing the necessary level of detail when needed.

Modeling-Specific Storytelling Techniques

When crafting a BPMN model using Process-bility, it's important to remember that the way elements are named, structured, and visually represented all contribute to the overall narrative of the process. The following modeling-specific techniques ensure that the story behind the process is conveyed clearly and effectively.

1. Verb + Object Naming Convention for Activities/Tasks, Sub-Processes, and Process Names

Technique:

Always name activities, tasks, sub-processes, and the overall process using the Verb + Object format (e.g., "Approve Invoice," "Send Quotation").

Fg8.5 Verb + Object

Reason:

This format emphasises action and object, making it immediately clear what the task. It eliminates ambiguity and focuses on what is being done.

This structure ensures that the focus is on What the activity accomplishes, without including unnecessary details such as how, why, who, or when. Clear activity names ensure better storytelling, with the action driving the process forward.

2. "Focus on the What" Concept:

Fg8.6 WHAT, not HOW

A stranded Man's Story

Imagine a man stranded on an island, desperate to attract the attention of a rescue helicopter flying above.
To get help, he builds a fire, hoping the smoke will catch the helicopter's attention. Now, the key objective here isn't about how he signals for help—it's about the what: getting attention to be rescued.

If we were to describe this scenario incorrectly by focusing on the how, we might say:

"Build a Fire" or "Create Smoke"

While these describe specific actions, they miss the larger point. The ultimate goal isn't just to make smoke — it's to attract attention and get help. Therefore, the correct naming should be:

"Get Attention for Rescue" or "Get Attention"

This aligns with the focus on the what concept: naming actions based on what they aim to achieve, not the exact method used. The man could signal using a flare, a mirror, or waving his arms—the goal remains the same:
get attention. This approach ensures the activity reflects the purpose, keeping it adaptable and clear.

3. Event in Noun + Past Participle

Fg8.7 Past Participle

Technique:
Use the Noun + Past Participle format for naming events (e.g., "Invoice Received," "Approval Granted") or just Noun.

Reason:
This structure emphasises that something has already happened or been completed. It conveys the state of the object at that point in the process. For example, "Invoice Sent" tells the reader that the invoice has already been sent, making it clear where the process is and what to expect next. This also fits into the storytelling aspect of process modeling, where events signal significant points in the narrative.

4. Avoid Role Name in Activity Names

Technique:
Avoid including roles in Activity/Task names (e.g., Hiring Manager review job appplication) , unless the role is not represented in a Swimlane.

Reason:
Roles are already represented visually in the Swimlanes, so including them in task names adds redundancy and clutter. The goal is to keep task names as clear and concise as possible. An exception can be made if the role is critical to understanding the task and there is no Swimlane for that role (e.g., "Submit Report to CEO" if the CEO does not have a Swimlane).

5. Meaningful Start and End Event Naming

Technique:
Always give Start and End Events meaningful names that describe the event's context. Avoid generic terms like "Start" or "End." Instead, follow the Noun + Past Participle format (e.g., "Order Placed," "Order Fulfilled").

Reason:
Descriptive names for Start and End Events provide immediate context and make it clear what triggers the process and what outcome is expected. This approach aligns with the storytelling principle, ensuring that each process has a clear beginning and conclusion. Additionally, when processes are linked across models, consistent naming helps maintain clarity and continuity.

6. Use of Intermediate Untype Events

Technique:
Use Untyped intermediate events to represent significant points in the process flow (e.g., "Quotation Validated").

Reason:
Events in BPMN are not just placeholders—they represent meaningful stages/outcome in the process journey. Intermediate Events help to break up the process and provide context for key outcomes of actions or changes. Using Untyped Events gives you flexibility in defining significant moments that are critical to storytelling but may not be tied to a specific BPMN event type like a message or timer.

7. Aligning Start and End Events Across Linked Process Models

Technique:

When designing BPMN models, it's essential to maintain clarity and continuity between linked processes. In many cases, the End Event of one process directly leads to the Start Event of another. To create a seamless connection, the names of these events should either match or clearly indicate a relationship between the processes. Here are the two scenarios:

Scenario 1: Exact Same Name for Start and End Events

In some cases, the End Event of one process is exactly the same as the Start Event of another process. This happens when the outcome of one process naturally feeds into the next without any transformation. For example, in a sales process, the End Event "Quotation Sent" could match the Start Event "Quotation Sent" in the delivery process. This shows that both processes are dealing with the same object and the same moment in time, ensuring seamless transition between them.

Fg8.8 Identical Start and End

Scenario 2: Different Names Reflecting Related Events

There are also situations where the End Event of one process and the Start Event of another are related but have different names, as they represent two sides of the same interaction. For instance, the End Event "Quotation Sent" in a sales process may match the Start Event "Quotation Received" in a procurement process. Though the names differ, they indicate a clear connection between the two processes, signaling the handoff of the same object or information.

Fg8.9 "Mirroring" Start and End

Reason:
When Start/End event names are consistent or clearly related, stakeholders can easily follow the flow from one process to another, reducing confusion and ensuring smoother transitions. This alignment also prevents gaps in understanding between different teams or departments, allowing them to see how their work fits into the larger process narrative. It's an essential storytelling technique that strengthens the connection between processes and helps maintain a cohesive business flow.

The Benefits of Storytelling in BPMN

By applying the principles of storytelling, your BPMN models will not only be more engaging but also far easier to understand. Process-bility ensures that every process has a clear narrative, with a well-defined main character, structured journey, and decomposed sub-processes where necessary.

This storytelling approach ensures that your BPMN models are more than just technical documents—they become powerful tools for communication, collaboration, and process improvement.

Closing Thoughts

Storytelling is the backbone of every successful BPMN model. It transforms a collection of activities, decisions, and events into a coherent process that stakeholders can easily follow. By identifying the main object of focus, crafting a clear narrative, and structuring the process to be digestible at every level, we ensure that our models speak to everyone, from business users to technical teams.

But understanding the principles of Storytelling is only the begining. In the next chapter, we'll dive deeper into how to apply both Process-structure and Modeling-Specific storytelling techniques in real-world BPMN models. You'll see practical examples of how Process-bility turns theory into action, creating models that communicate with clarity while maintaining the integrity of the business process. Let's explore how to bring these stories to life in your process models.

Chapter 9

Storytelling In Actions

- Case 1: Transforming a Citizenship Application Process
- Case 2: Crafting The Journey - From Level 3 to Level 4 Process Views
- Closing Thoughts

> **Do not model business processes with the intent to demonstrate your skill in modeling, but with the intent to provide clarity and insights to the business.**

Case 1: Transforming a Citizenship Application Process

In this section, we'll demonstrate how storytelling principles can enhance BPMN models using a real-world scenario: the citizenship application process. The original model presents a detailed, low-level view of the process. While it accurately captures the sequence of tasks, it can overwhelm stakeholders with too many details at once.

An Overwhelming Low-Level Process Model

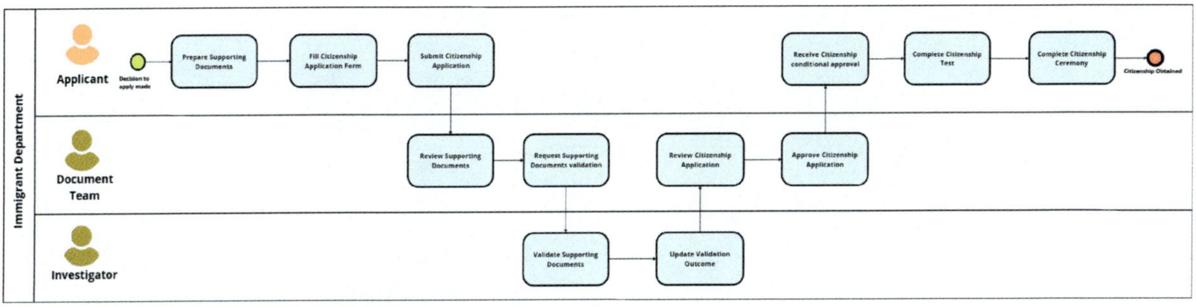

Fg9.1 Super detailed process model

Imagine if the actual business process had twice the number of activities—the sheer amount of information would make the model difficult to read and understand. In situations like this, storytelling through BPMN modeling becomes essential. It's not just about showing every task but presenting the process clearly and logically so that stakeholders can engage with it, understand it, and take action.

The Challenge:

- Too many low-level details clutter the flow.
- Hard to identify key milestones and responsibilities.
- Overwhelms stakeholders with unnecessary complexity.

Applying Decomposition Storytelling Techniques

Now that we've identified the challenges, let's transform the model using decomposition and storytelling principles. With sub-processes and a cleaner structure, the model becomes easier to follow and more focused.

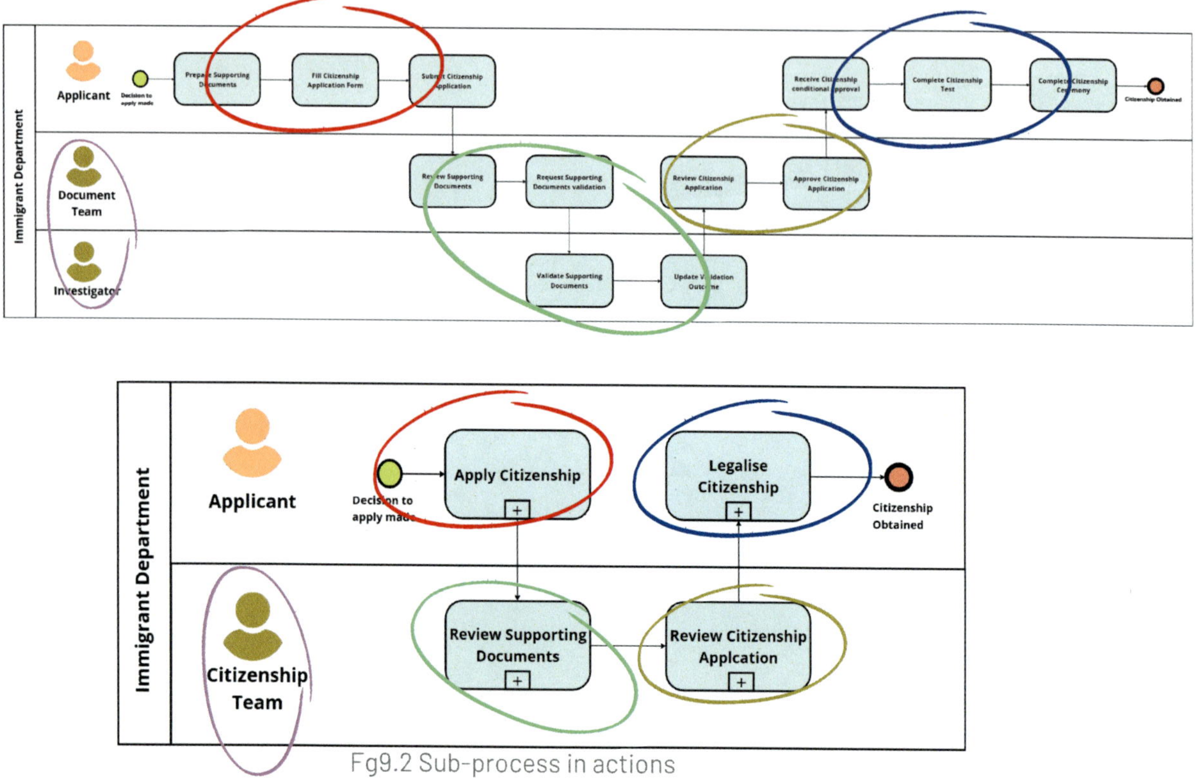

Fg9.2 Sub-process in actions

The Changes:

- Merging Document Team and Investigator as Citizenship Team
- Group Activities into four meaningful sub-processes name
    - Apply Citizenship
    - Review Supporting Documents
    - Review Citizenship Application
    - Legalise Citizenship

This way, the sub-process names act as high-level summaries that allow stakeholders to focus on intent and outcomes, rather than getting lost in details.

Key Improvements

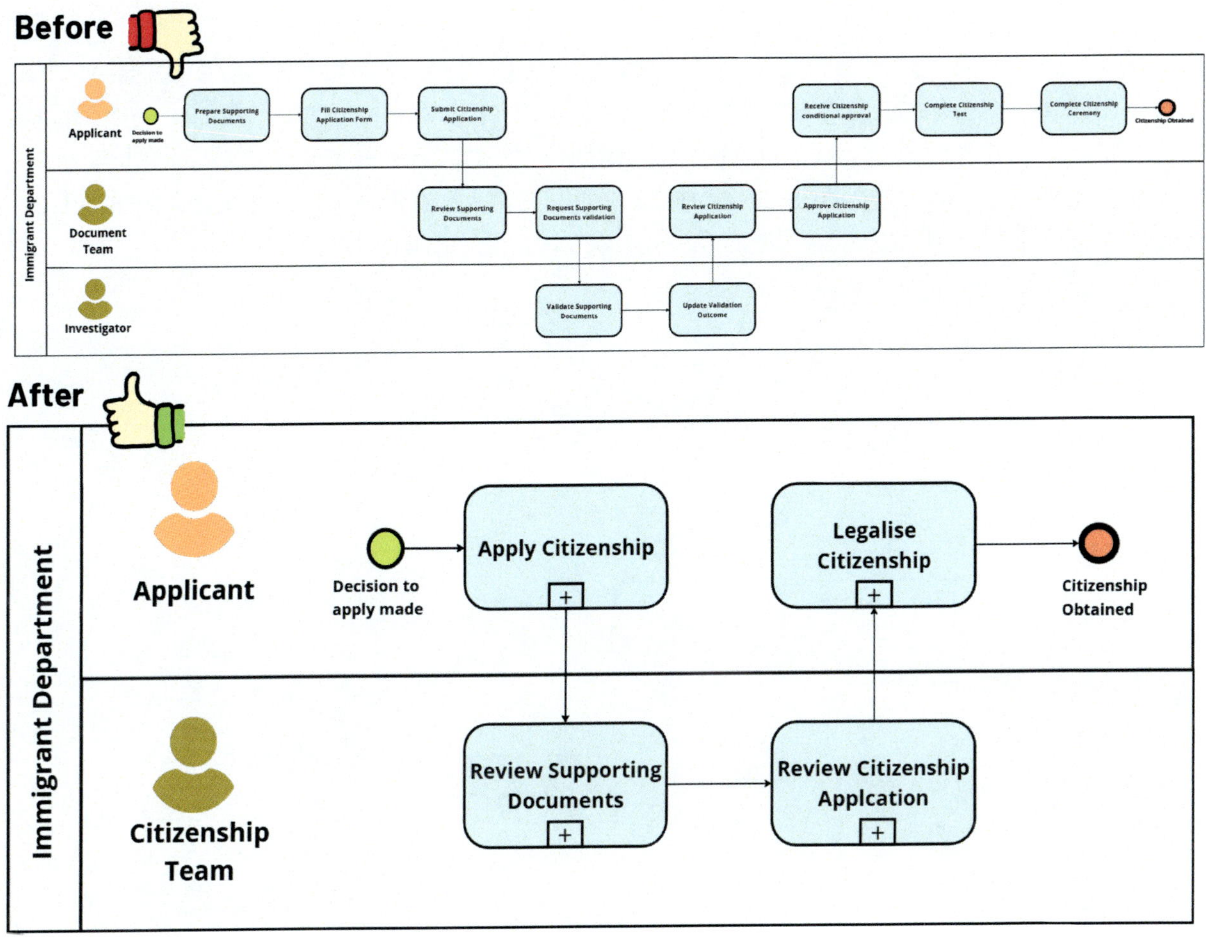

Fg9.3 Sub-process in actions - Before and After

Here are the key differences between the two versions:

- Decomposition into Sub-Processes: Focuses on key phases instead of granular tasks.
- Clearer Flow: Participants can easily follow the handoffs and responsibilities.
- Improved Usability: The model serves as an effective communication tool, not just documentation.
- Drill-Down Capability: Readers can drill down into the sub-processes to explore detailed tasks only when needed, maintaining clarity while offering full transparency.

Joining the Stories

The End Event of one process and the Start Event of another are related but have different names, as they represent two sides of the same interaction. End Event "Citizenship Application submitted" of Sub-Process Submit Citizenship Application has becomes the Start Event "Citizenship Application received" of Sub-Process Review Supporting Documents

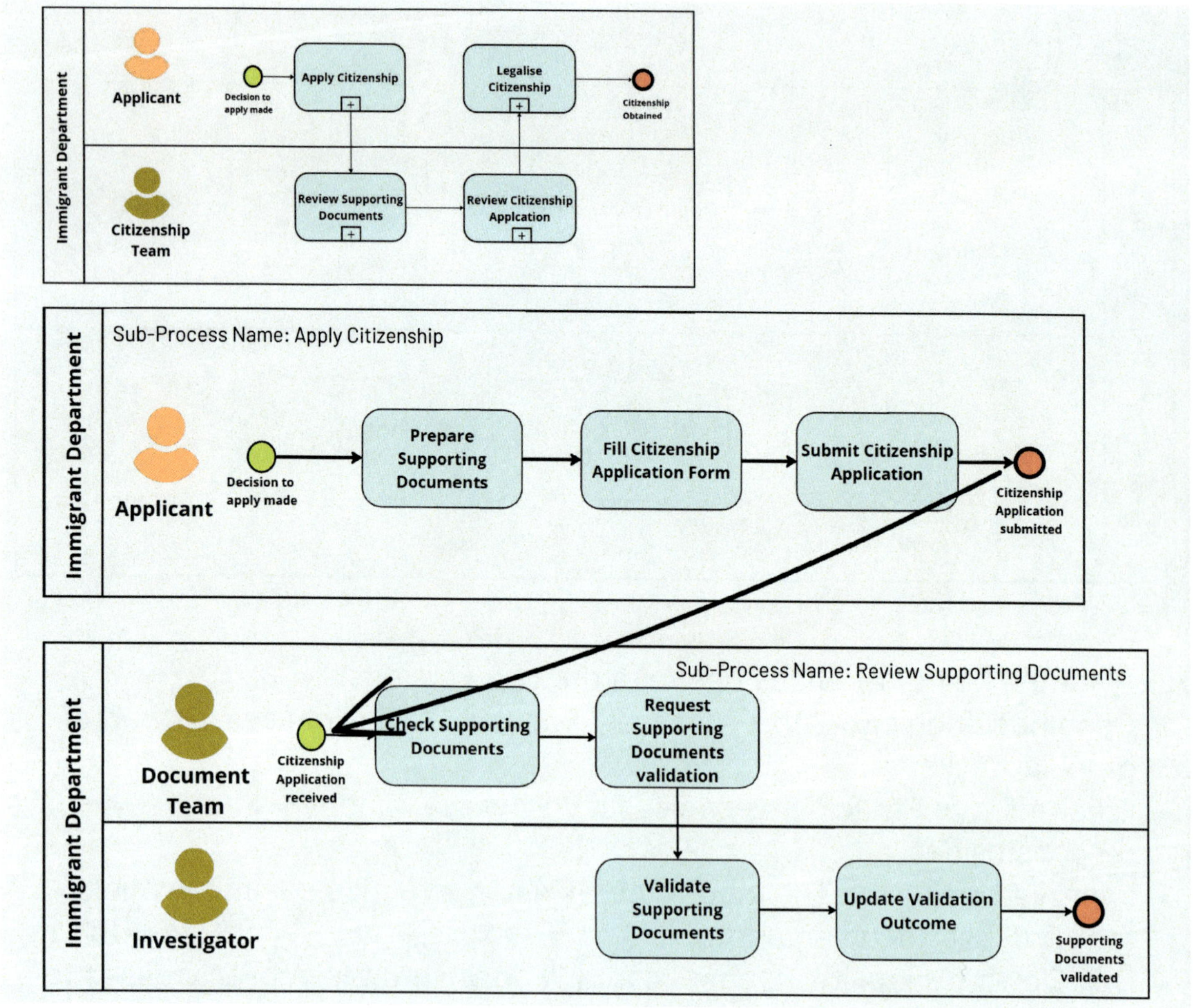

Fg9.4 Sub-process in actions - Start/End Events naming

Case 2: Crafting The Journey - From Level 3 to Level 4 Process Views

Let's explore the solar panel installation process using a multi-level BPMN approach. This example demonstrates how storytelling through BPMN can help present a clear, structured process by breaking down complexity into meaningful levels.

Level 3: The High-Level Process View – "Provide Solar Panel Installation"

At Level 3, we present a high-level process model showing the happy path —the ideal scenario where everything goes smoothly. This level uses sub-processes to focus on the core phases of the installation process, giving stakeholders a quick overview.

What does the Level 3 model highlight?

- The key stages from enquiry to installation completion, without diving into excessive detail.
- A clean, streamlined overview of the entire process, keeping the narrative focused on the positive flow.
- Each phase (like capacity assessment and installation) is grouped as a sub-process, providing a clear summary.

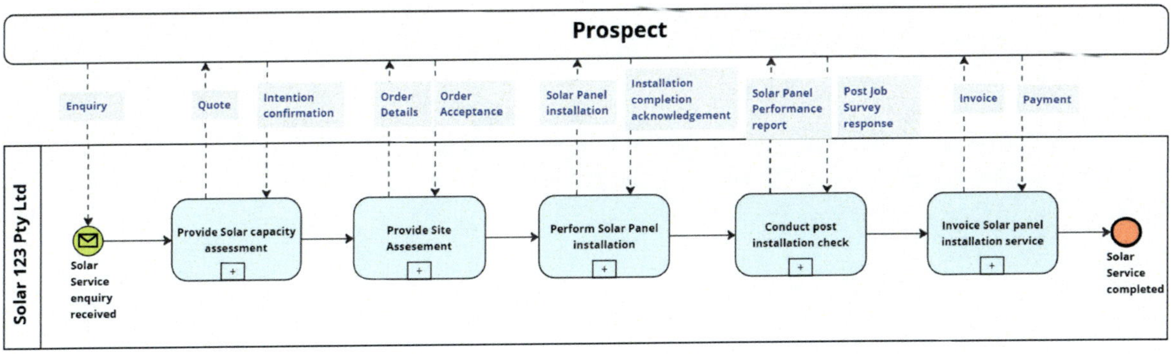

Fg9.6 End to End BPMN Model (Level 3)

Drilling into Details: Level 4 Process View of "Provide Solar Capacity Assessment"

At Level 4, we zoom into the first sub-process, "Provide Solar Capacity Assessment." This layer reveals the specific variations and different paths based on customer type (residential vs. commercial). It also introduces the not-so-happy paths to reflect real-world scenarios, such as:

- Rejected Quotes: The prospect may reject the quote, requiring follow-ups or amendments.
- Different Assessments: The process varies depending on whether the customer is residential or commercial, showcasing the flexibility needed.

What does the Level 4 model show?

- Detailed steps within the capacity assessment process.
- Variations and alternate paths, such as different assessments for residential and commercial customers.
- Realistic outcomes, including rejected quotes and follow-ups, providing a complete view of the process.

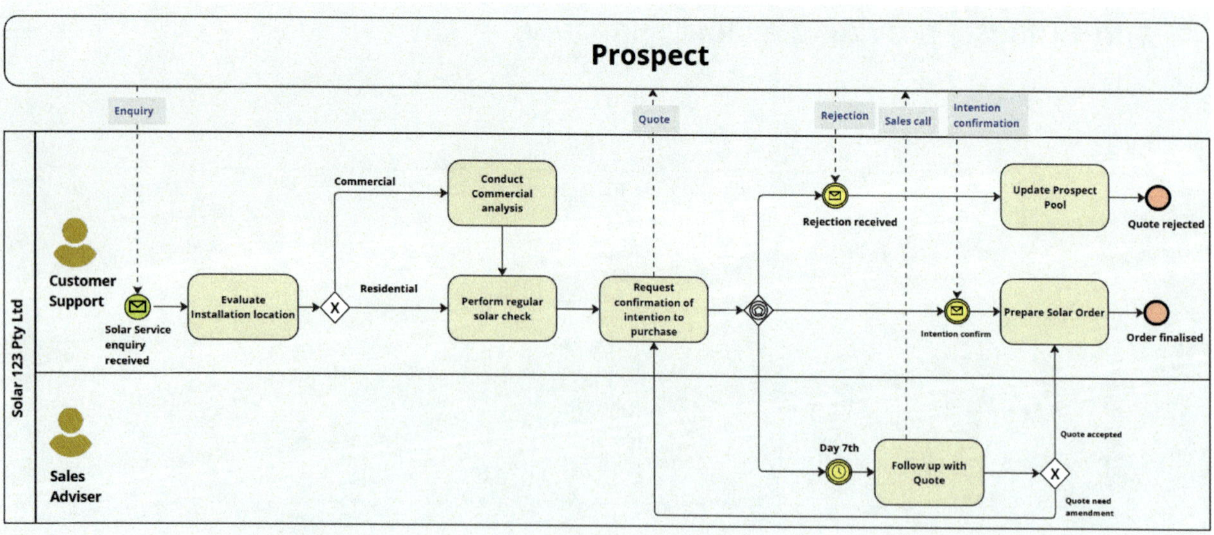

Fg9.7 BPMN Model Level 4

Clarity Through Multi-Level Storytelling

By layering the process across multiple levels, we maintain both clarity and detail:

- Level 3 offers a big-picture view, helping stakeholders quickly grasp the key phases.
- Level 4 allows for deeper exploration, showing the full complexity only when needed.

This storytelling technique ensures that the process remains engaging and usable for different audiences—those looking for a high-level summary and those needing detailed insights.

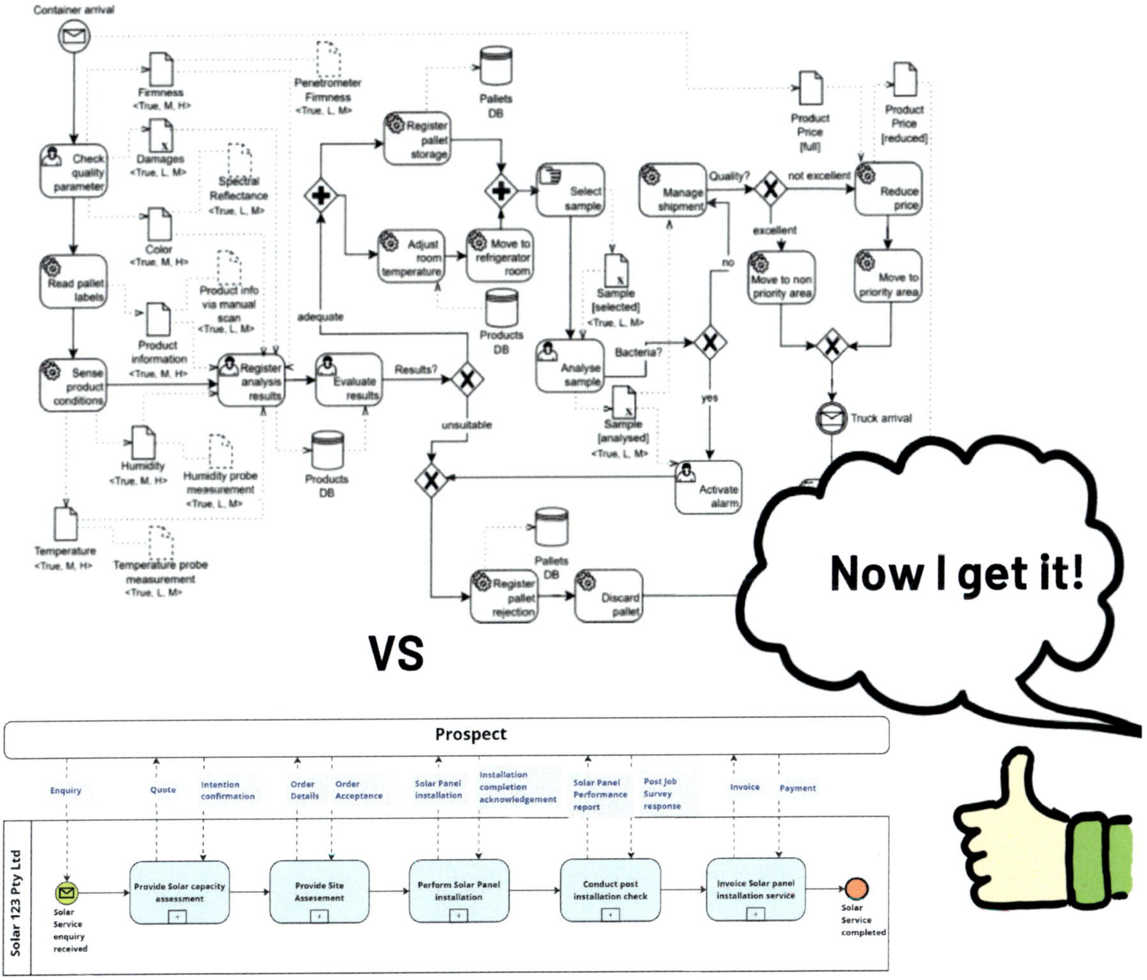

Fg9.8 Now I get it! Storytelling with sub-processes

Closing Thoughts

At the core of every impactful BPMN model is a well-told story. Storytelling is the top priority—it ensures the model not only documents processes but also communicates meaning, aligns teams, and drives action. Clear narratives help stakeholders understand their roles and stay engaged.

Visual Balance and Simplicity support Storytelling by keeping the model structured and focused. Most of the time, these principles work seamlessly together, but at times, the story takes precedence, requiring trade-offs.

Ultimately, effective communication comes first — the goal is to create models that foster understanding and action. Visual Balance and Simplicity enhance the story, not overshadow it.

As we conclude this section, the focus shifts back to you—how will you use these principles in your own work? Remember that every process tells a story. Your challenge is to tell it well — clearly, meaningfully, and with purpose.

Q A

Does Storytelling, Visual Balance, and Simplicity work hand in hand?

Storytelling, Visual Balance, and Simplicity are the three core pillars of Process-bility, and they work best together to create BPMN models that communicate effectively.

Storytelling is the heart of the model—it's what gives the process meaning and relevance. It turns a series of tasks into a narrative that stakeholders can understand, follow, and connect with. Storytelling ensures the model isn't just a diagram; it's a way to engage and align people around a common goal.

Visual Balance supports Storytelling by making the model easy on the eyes, structured, and intuitive. When a model has good Visual Balance, it guides the reader's focus naturally, drawing attention to key parts of the process without overwhelming them. Visual Balance reduces visual noise, so the story stands out.

Simplicity, on the other hand, ensures that the model is focused and free of unnecessary details. It's about using only what's essential to communicate the process effectively. When we simplify, we make the process clearer and faster to grasp, making it easier for stakeholders to take action.
(continued....)

Q A

Does Storytelling, Visual Balance, and Simplicity work hand in hand? (continued)

Most of the time, these three elements work seamlessly together. But sometimes, for the sake of telling a clearer story, you may need to compromise on Visual Balance and/or Simplicity. The key is always to remember that the ultimate goal of Process-bility is effective communication. A great model isn't just technically correct—it's a model that people understand and want to use.

In Process-bility, these elements are like a team. Each brings its unique strength, but together, they amplify each other to make your BPMN model an invaluable tool for achieving meaningful business outcomes.

Chapter 10

Correlating Agile User Stories with BPMN Models

- Introduction : Connecting Stories with BPMN
- The Problem with User Stories Alone
- How Process-Bility Aligns User Stories and BPMN Models
- User Stories and BPMN Process Models in Action

Introduction: Connecting Stories with BPMN

Agile user stories focus on delivering value through small, incremental features. BPMN models, on the other hand, describe the flow of activities and handoffs in a structured process. The challenge lies in bridging these two perspectives — connecting the narrative of user needs with the structured workflows that support them. Process-bility offers a way to bring both together, ensuring alignment between user expectations and the underlying process logic.

The Problem with User Stories Alone

User stories tell what needs to be done from the end-user perspective, but they often lack clarity on the "bigger picture"—especially when multiple teams, systems, or steps are involved. Without a clear process view, user stories may:

- Miss critical dependencies.
- Lead to misaligned expectations across teams.
- Cause delays when handoffs between roles are unclear.

How Process-Bility Aligns User Stories and BPMN Models

Process-Bility ensures that each user story is linked to a process flow, providing both narrative clarity and operational structure. Here's how:

Decomposition into Sub-Processes

Break down larger user stories into smaller tasks, mapped to BPMN sub-processes. This allows teams to see the process behind the story and understand who is responsible for what.

Role Clarity with Visual Balance:

Use person icons in Swimlanes to clearly indicate roles responsible for each task, ensuring smooth handoffs.

Storytelling with the Flow:

A well-modeled BPMN process supports the narrative in user stories by showing the actual flow—how activities connect to achieve the desired outcome.

User Stories and Process Models in Action

In agile development, Themes, Epics and User Stories serve as a structured approach to capturing requirements and breaking down work in manageable increments. However, the journey doesn't end with defining these elements. To turn them into actionable insights, we link them to process models, which visually depict the flow of actions, decisions, and roles involved. Here's a look at how a Theme can cascade into Epics and User Stories that, in turn, map directly into BPMN process models.

Theme: Employee Self-Service Portal

Our example starts with a high-level Theme: Employee Self-Service Portal. This theme is focused on empowering employees to manage their own HR-related activities through a digital portal. By allowing employees to initiate actions independently, the portal reduces HR workload, increases transparency, and improves the employee experience. The Employee Self-Service Portal theme encompasses various HR-related functionalities, one of which is Leave Management.

Epic: Leave Management

Within the Employee Self-Service Portal theme, Leave Management is a critical Epic. This Epic includes all necessary features for an employee to apply for leave, a manager to review and approve or deny the request, and HR to update records accordingly. The Leave Management Epic provides an end-to-end structure, from request submission to leave approval, balancing user autonomy with organizational oversight.

By breaking down the Leave Management Epic into User Stories, we can address specific functionalities and align each story with the corresponding steps in a process model, ensuring that each requirement is actionable and traceable.

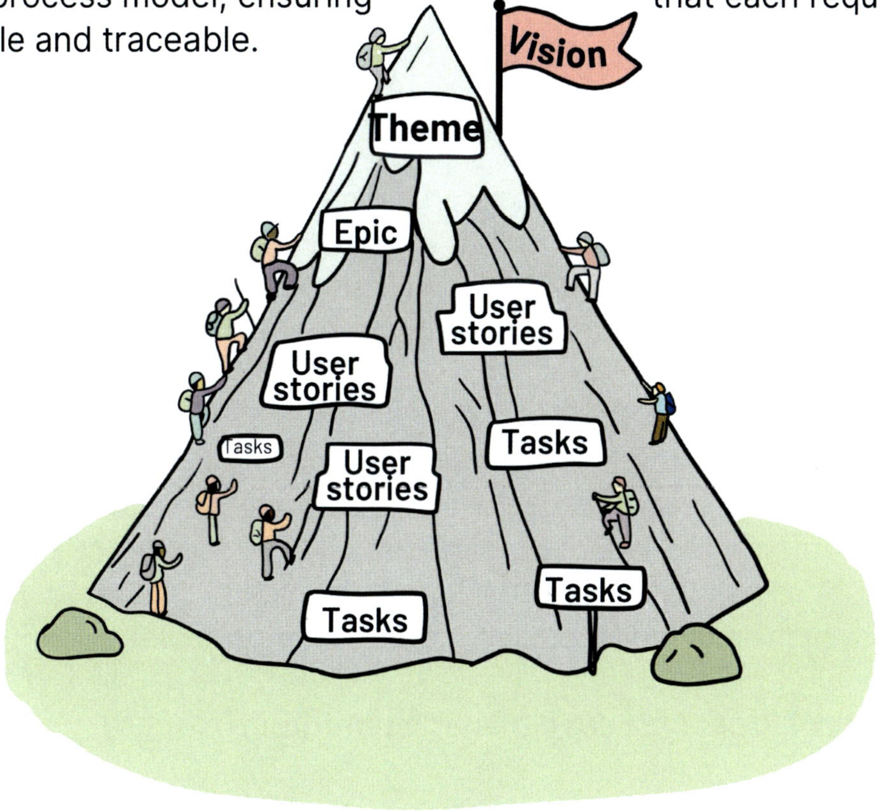

Fg10.1 Mountain climbing : Tasks to Vision

User Stories within the Leave Management Epic

To capture the details of the Leave Management process, we define individual User Stories that represent discrete user needs within this Epic:

1. Employee Leave Application
   - As an employee, I want to apply for leave so that I can request time off through the portal.
2. Manager Approval
   - As a manager, I want to review and approve or deny leave requests from my team so that I can ensure adequate coverage during any period.
3. HR Leave Policy Configuration
   - As an HR administrator, I want to set up leave policies in the system so that leave accruals, carryovers, and approvals are managed according to company rules.

Each of these user stories focuses on a distinct part of the leave management process, ensuring that every functional requirement is captured. With user stories defined, we can now map them into BPMN process models to show how these requirements translate into business workflows.

Linking User Stories to BPMN Process Models

Each User Story in the Leave Management Epic maps directly into tasks, decisions, or roles within the BPMN model, which is split into two levels to cover both high-level and detailed views of the process.

Level 3 BPMN Process: Leave Management (End-to-End)

The Level 3 BPMN process model provides a high-level overview of the Leave Management process, aligning with the Epic's scope. It captures the entire journey from the initial leave request to the final record update, giving stakeholders a clear understanding of the main stages without overwhelming detail.

Key Stages in the Level 3 Process Model:

1. Initiate Leave Request – Linked to Employee Leave Application
2. Review Leave Request – Linked to Manager Approval
3. Configure Leave Policy – Linked to HR Leave Policy Configuration

These stages align with the core actions required to fulfill the Leave Management Epic, translating high-level requirements into an end-to-end visual flow.

Level 4 BPMN Process: Detailed Leave Management Workflow

The Level 4 process model goes deeper into each stage, showing the step-by-step details that align with individual User Stories. This level of detail allows developers and stakeholders to understand the specific tasks, decision points, and role-based responsibilities.

Detailed Process Steps in the Level 4 Model:

1. Initiate Leave Request:
    - Employee selects leave type and dates, checks available balance, and submits the request (linked to Employee Leave Application).
2. Manager Review and Decision:
    - Manager reviews the request, checks the team schedule, and either approves or denies the leave (linked to Manager Approval).
3. HR Configure Policy:
    - HR updates leave policy as needed.

The Level 4 process model ensures that each User Story is fully represented in the workflow, allowing for a clear and accurate execution of the leave management process.

Outcome of Linking User Stories to Process Models

By linking User Stories to specific tasks and decisions within the BPMN model, we ensure that each requirement is operationalised within a visual workflow. This approach provides several benefits:

- Traceability: Each User Story's requirements are traceable within the process model, making it easy to verify that the final implementation meets user needs.
- Clarity: The high-level (Level 3) and detailed (Level 4) models provide clarity for stakeholders at different levels, supporting effective communication and alignment.
- Actionable Insights: Process models turn requirements into actionable steps, ensuring that the final workflows are both user-friendly and compliant with organizational policies.

This alignment of Themes, Epics, User Stories, and BPMN Models demonstrates how agile requirements can be effectively operationalised, from high-level goals down to individual tasks, driving real business value and delivering a seamless user experience.

Visualising Epics and User Stories in BPMN Models

This page illustrates an example of a Level 3 BPMN End-to-End Model alongside a Level 4 BPMN Model. The Level 3 model provides a high-level view of the Leave Management Epic, showing the end-to-end process from request initiation to approval and policy update. Meanwhile, the Level 4 model delves into the detailed steps for the Leave Application User Story, depicting each action an employee takes when submitting a leave request, along with validations and decision points.

Level 3 BPMN Process: Leave Management (End-to-End) | EPIC

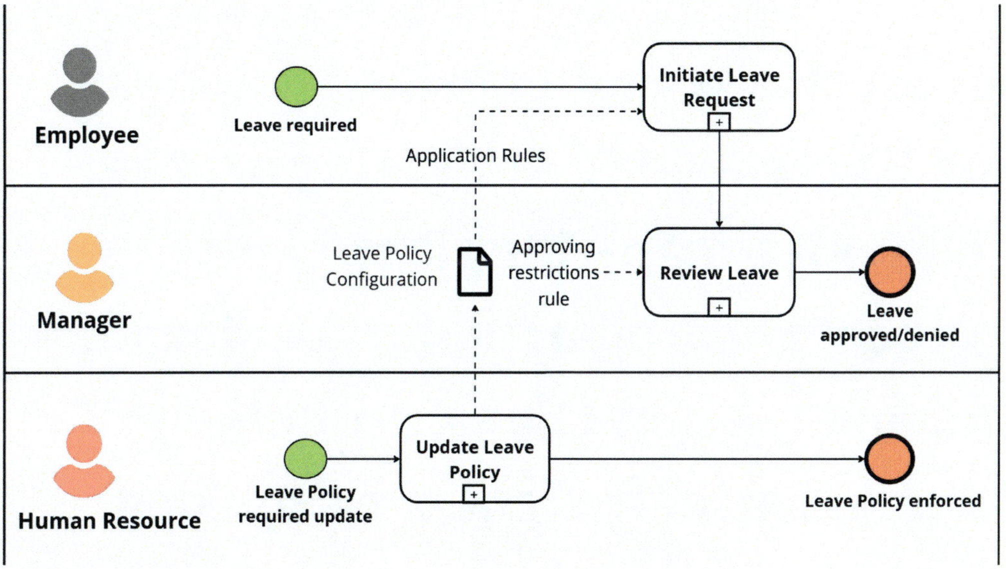

Level 4 BPMN Process: Initiate Leave Request : User Story

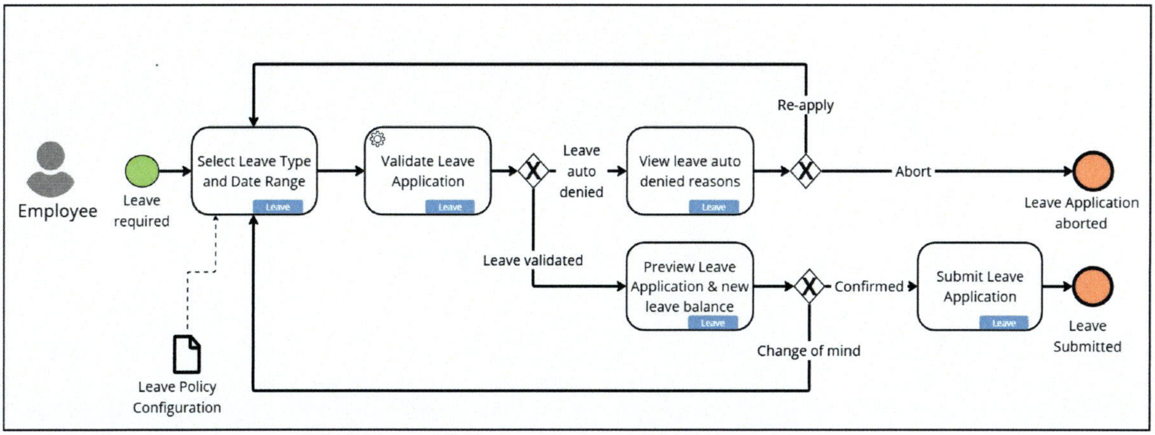

Fg10.2 EPIC-User Story to BPMN Level 3 End to End and BPMN Level 4

User Stories without business process models are akin to navigating a maze without a map - they lack clarity, direction, and a clear understanding of the overall objectives, potentially resulting in inefficient or misguided solutions.

Chapter 11

Real-World Business Scenarios Modelling Challenges

Introduction

In this chapter, we dive into practical applications of Process-bility principles, tackling real-life business problems through clear, actionable BPMN models. Each case presents a common challenge and explores how effective process modeling can simplify complexity, enhance operational flow, and achieve meaningful results.

From managing procurement processes to handling customer service workflows, these scenarios are crafted to provide a practical toolkit. Use them as inspiration and guidance to address similar challenges in your own organization, applying Process-bility's focus on Storytelling, Visual Balance and Simplicity to create models that communicate.

BPMN Challenge Scenario 1: Quote Evaluation Process

Imagine you're working as a Business Analyst, tasked with modeling the quote evaluation process for a company sourcing a product or service. Here's the setup:

The company must send an Invitation to Quote to at least three providers. The quote evaluation process can begin under one of the following conditions:

- After the closing due date, regardless of how many quotes are received.
- Immediately, if more than five quotes are received before the closing due date.

Your Challenge:

Create a BPMN model for this process using the principles of Process-bility— Storytelling, Visual Balance and Simplicity. Your model should clearly represent the conditions that trigger the evaluation phase and simplify the decision-making process.

BPMN Challenge Scenario 1: (Continued)

Questions to Consider:

- Decision Flow: How would you handle the decision flow to start the evaluation?
- Gateway Choice: Would you use an exclusive gateway or another type of gateway to manage the conditions?
- Clarity and Communication: How do you ensure the process is clearly communicated to stakeholders, especially regarding the timing and conditions for starting the evaluation?

Take a moment to think through these questions and model the process. On the next page, we'll explore a typical BPMN model and a Process-bility approach to see how each handles the challenges of clarity and simplicity.

Good luck!

BPMN Challenge Scenario 1: (Continued)

Take a few moments to model the scenario on your own. Remember, the best way to truly learn is by doing. Give it a shot—embrace the Process-bility principles you've learned so far.

When you're ready (or if you really, really can't wait), head over to next page to see the Process-bility approach to this challenge.

Good luck, and have fun with it! You got this!

> **How we model is equally, if not more important than what we model.**

BPMN Challenge Scenario 1: Process-bility Model Answer

Process Name: Obtain Quotes

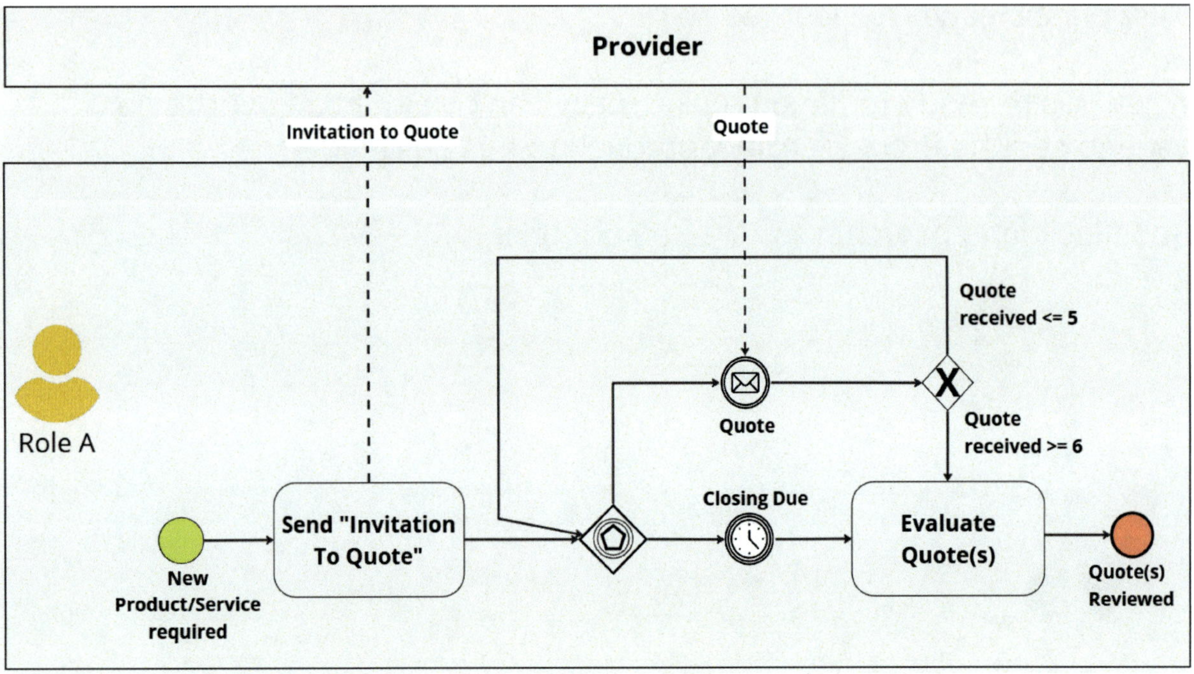

Fg11.1 Model answer for Challenge #1

BPMN Challenge Scenario 2: Expense Report Approval Process

Imagine you're tasked with modeling the following business process:

1. Submission and Approval Needed for Payment:

The Employee submits an expense report to the Manager for approval. Before any payment can be made, the report must receive approval from the Manager.

2. Manager's Rejection:

The Manager has the authority to reject an expense report if it doesn't meet necessary criteria.

3. After Rejection:

If rejected, the Employee must address the issues and resubmit the report for approval.

4. Resubmission:

The resubmitted report goes through the approval process again, and only after it's approved can Finance proceed with payment.

5. Additional Approvers for High Amounts:

For higher-value reports, up to two additional sequential approvals from higher-level managers may be required before payment.

Use your Process-bility knowledge to model this process in BPMN.

- How will you represent the decision flows for Manager's approval or rejection?
- If additional approvals are needed, how would you illustrate this condition without overcomplicating the model?
- How will you keep the flow clean and understandable, especially with potential resubmissions?

BPMN Challenge Scenario 2:(Continued)

Gotcha trying to sneak a peek, didn't I?

Hold on there! The answer will be there, but first, let's put your skills to the test. Take a few minutes to tackle the scenario yourself. After all, the best learning comes from giving it a go.

Think about the Process-bility principles you've learned—Simplicity, Storytelling, and Visual Balance. You've got this!

Tip: When modeling this, avoid creating a separate wimlane for each additional approver. Imagine if the business rule allows up to 5 extra approvals—that wouldn't be sustainable (or simple)! Instead, think of a scalable way to represent this approval flow.

When you're ready (or if the suspense is just too much), head over to the next page to compare your model with the Process-bility solution.

Good luck, and enjoy the challenge!

BPMN Challenge Scenario 2: Process-bility Model Answer

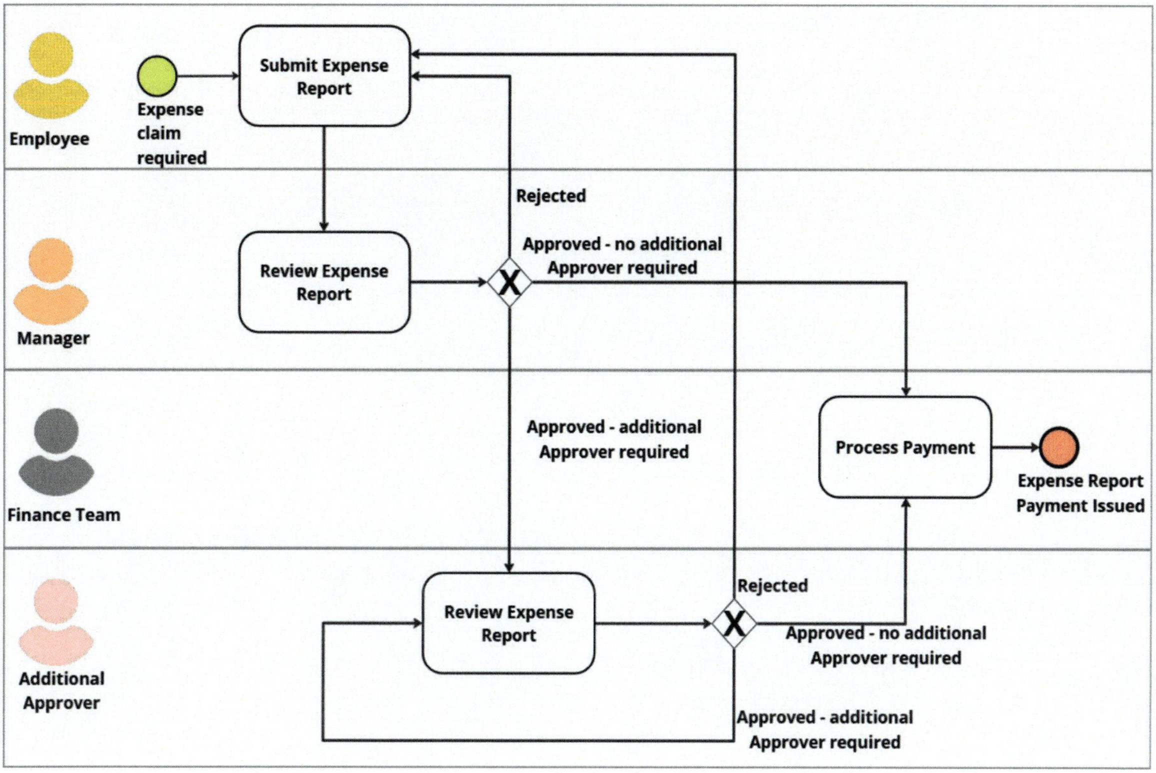

Fg11.2 Model answer for Challenge #2

What do you think of the model answer? Can you propose an alternative way to model this scenario? Do you think the 'cross-over' is acceptable, or could it be avoided? Perhaps you can simplify it further—maybe even reduce the number of Swimlanes? Remember, the ultimate goal is not just to model accurately but to embrace Simplicity, Storytelling, and Visual Balance.

Give it a shot and email your model to me at kevin.tan@process-bility.com

I'd love to see your approach!

(Or flip to the next page to view an alternative model.)

BPMN Challenge Scenario 2: Process-bility Alternate Model Answer

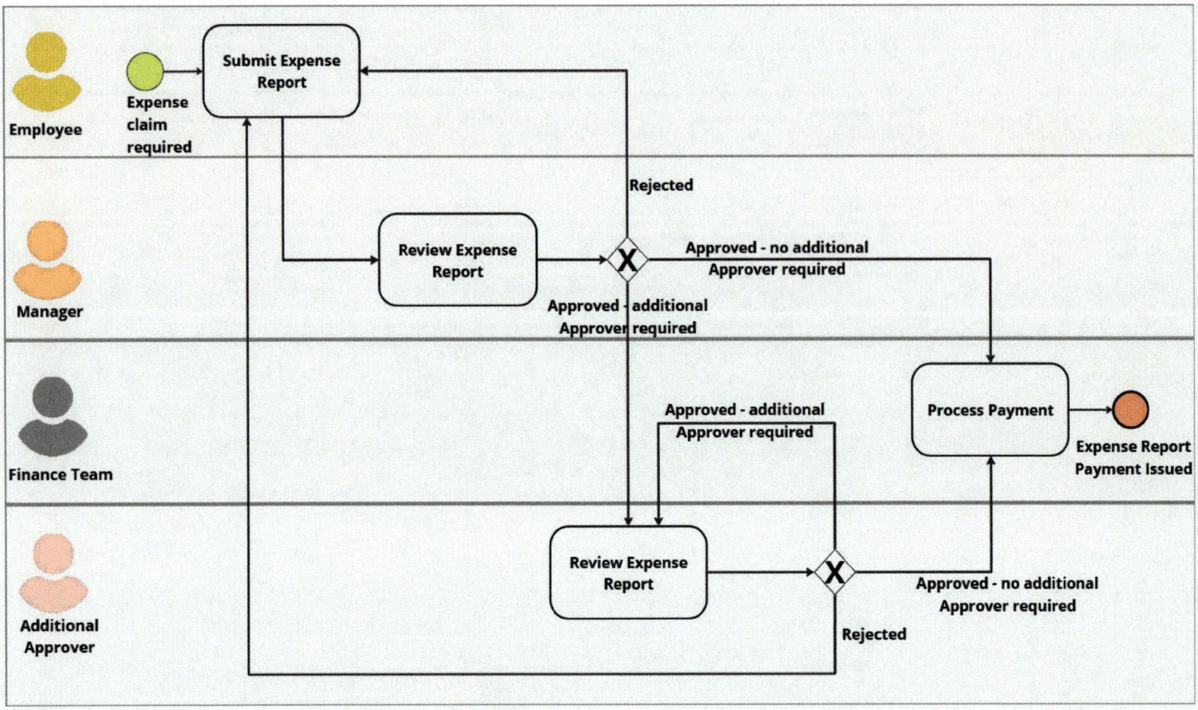

Fg11.3 Alternate Model answer for Challenge #2

BPMN Challenge Scenario 3: Customer Refund Process

As we wrap up this chapter, I invite you to apply what you've learned and tackle a final Business Scenario Challenge. This is your opportunity to put Process-bility principles—Storytelling, Visual Balance and Simplicity—into action.

Model the scenario using your newfound knowledge, and feel free to email your BPMN model to me at Kevin.Tan@process-bility.com. I'll personally review each submission, and selected answers will be featured in future updates of this eBook, with your name credited as a contributor.

Let's keep the journey going—learning, improving, and making process modeling simpler and more impactful, one challenge at a time!

Let's have fun

Imagine you're tasked with creating a BPMN model to manage customer refunds, involving multiple steps and decision points to ensure a smooth and efficient workflow.

1. **Daily Report Download:**

   Each day, the Customer Service team downloads a refund request report from the CRM system into an Excel spreadsheet. This report lists all pending refund requests submitted by customers.

2. **Initial Eligibility Assessment:**

   Customer Service first checks if the product qualifies for return based on:
   - Time Since Purchase: Eligible if within 13 months of purchase.
   - Reason for Return: The reason provided by the customer, such as a defect or a change of mind.

BPMN Challenge Scenario 3: Customer Refund Process

3. Product Condition Check by Warehouse:

For returns meeting initial criteria, the product is sent to the warehouse for inspection:

- Defective Product:
  - If the defect is found within 12 months of purchase, the customer qualifies for a full refund.
  - If between 12 and 13 months, the customer is offered credit for a future purchase.
- Good Condition, Change of Mind:
  - If the product is in good condition but returned due to a change of mind, a partial refund is processed after deducting restocking fees.
- Damaged Beyond Eligibility:
  - If the product is too damaged, a refund is not possible.

4. Approval and Processing by Finance:

For eligible refunds, Finance reviews the refund amount and approves based on the warehouse's assessment. If any discrepancies arise, Finance may seek clarification from Customer Service before completing the payment process.

Chapter 12

Designing Process Architecture

- **Introduction**
- **A Hybrid and Practical Approach with Process-bility**
- **Process Guardian in Process-bility**
- **Closing Thoughts**

Introduction

Processes are the foundation of every successful organization. Without a clear, structured approach to process architecture, inefficiencies grow, silos remain unchallenged, and business performance stagnates. This chapter will introduce you to the Process-bility approach to designing process structures that balance functional expertise with end-to-end efficiency.

In this chapter, you will step into the role of a Business Process Architect for Coles Group, where you will be tasked with structuring their processes in a way that fosters collaboration and operational clarity. You will learn how to define L1 to L4 processes, ensuring that business processes are documented, understood, and continuously improved.
By the end of this chapter, you will:

- Understand the importance of structured process hierarchy (L1 to L4).
- Learn how to design a practical process structure that balances functional expertise with cross-functional collaboration.
- See how Process-bility enables efficient and scalable process documentation.
- Explore the role of a Process Owner (or Process Steward, as I think is a more appropriate name, given the responsibility Process-bility design for this role) and how it differs from traditional governance models.

Let's dive in!

Welcome to Your Challenge

Imagine this: Coles Group, one of Australia's largest retailers, is facing increasing pressure to reduce costs, improve efficiency, and enhance customer satisfaction in an increasingly competitive market. While they have strong market presence, their internal processes have grown complex and fragmented over time.

Senior management has identified key challenges:

- Lack of process standardisation across different stores and departments.
- Inefficiencies in operations, customer service, and financial processes.
- No clear ownership of end-to-end processes, leading to misalignment between departments.
- Process documentation is outdated or underutilised, making it difficult for teams to improve, train, or audit processes effectively.

Now, you have been hired as a Business Process Architect to help Coles design a clear, structured, and scalable process architecture that aligns with their business strategy.

Fg12.1 Coles Group

Your Task

Your mission is to define the L1–L4 process structure for Coles Group, bringing clarity, consistency, and efficiency to their operations. You will work alongside department managers and business teams to structure processes in a way that fosters collaboration while maintaining functional expertise.

For this exercise, we will focus on four key functional areas that play a crucial role in Coles' business operations:

1. Operations (Core function)
2. Customer Service (Core function)
3. Human Resources (Support function)
4. Finance (Support function)

(Note: This is an imaginary case study. While Coles Group operates many more functional groups, we will simplify this example to four areas for clarity and learning purposes.)

Let's get started.

Step 1: Understanding the L1–L4 Structure

Before jumping into the specifics, let's break down the hierarchy:

- Level 1 (L1): Functional Groups – The highest level of business functions.
- Level 2 (L2): Capability Groups – The core operational capabilities that enable each function.
- Level 3 (L3): End-to-End Processes – The actual business processes that teams execute, often spanning multiple L1 and L2 groups.
- Level 4 (L4): Business Processes – More detailed process flows within each end-to-end process that define execution at a refined level.

A Hybrid and Practical Approach with Process-bility

Process-bility takes a hybrid approach to structuring business processes. L1 and L2 follow a more traditional functional structure, allowing organizations to manage expertise and responsibilities within business units. However, at L3, the approach shifts to an end-to-end process view, ensuring cross-functional collaboration and flexibility.

The start and end of an L3 process can be defined by the business as needed, rather than being restricted to predefined departmental silos. Using this approach ensures clarity across the organization, aligns teams, and removes process inefficiencies—a key challenge Coles must overcome.

Step 2: Defining Coles Group's Process Architecture

Level 1 (L1) – Functional Groups

To tackle Coles' process inefficiencies, the first step is to establish a structured L1 functional group hierarchy. For this exercise, we are focusing on four critical areas:

1. Retails Operations
2. Customer Service
3. Human Resources
4. Finance

These four groups have been selected as they directly impact cost reduction, customer satisfaction, workforce management, and financial sustainability—key areas that must be optimised to support Coles' business goals.

Level 2 (L2) – Capability Groups

Now, let's define the key capabilities within each function:

1. Retails Operations
   - Store Management
   - Inventory Control
   - Online Order Processing

2. Customer Service
   - In-Store Support
   - Loyalty & Rewards
   - Customer Feedback & Issue Resolution
   - Online & Call Centre Support

3. Human Resources
- Talent Acquisition
- Training & Development
- Worksafe & Compliance
- Payroll & Benefits Management

4. Finance
- Payment Processing
- Financial Reporting
- Budgeting & Forecasting
- Revenue & Expense

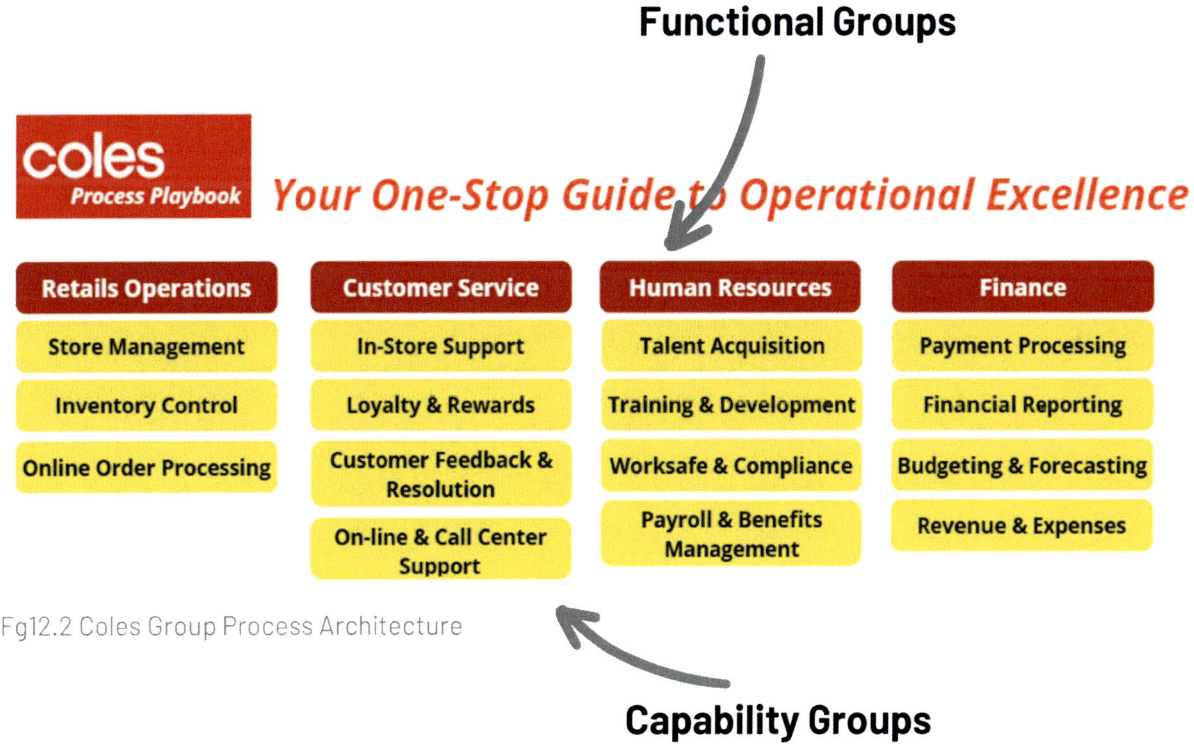

Fg12.2 Coles Group Process Architecture

Level 3 (L3) – End-to-End Processes (Cross-Functional & Cross-Capability)

For the Customer Service – In-Store Support capability, here are three key End-to-End processes:

1. Customer Refund & Resolution – Handling refund requests, verifying eligibility, processing refunds, and updating inventory.
2. Customer Complaint Handling – Managing customer complaints, logging issues, escalating where necessary, and ensuring resolutions meet service standards.
3. Loyalty Programme Enrolment & Support – Assisting customers with loyalty programme sign-ups, resolving issues with rewards, and ensuring accurate points allocation.

We will now focus on the first example: Customer Refund & Resolution as a detailed L3 process.

Business Scenario: In-Store Order Refund & Customer Resolution

Context:

A customer walks into a Coles retail store to request a refund for a recently purchased product. The process involves multiple functions: Customer Service (store staff), Operations (store manager), and Finance (cashier/payment processing team).

Process Description:

1. Customer Requests Refund (Customer Service)
    - The customer presents the product and receipt to the store staff.
    - Store staff logs the request in the system.
2. Refund Eligibility Check (Customer Service)
    - Store staff verifies the refund conditions:
        - Within the refund policy period?
        - Product in acceptable condition?
        - Receipt available?
    - If eligible, proceed to product return.
    - If not eligible, explain to the customer and reject the request.
3. Product Return Processing (Retails Operations)
    - Store manager inspects and approves the product return.
    - The returned item is recorded in inventory as "returned stock."
4. Refund Transaction Processing (Finance)
    - The refund is processed based on the original payment method:
        - Credit Card Purchase → Refund to the card.
        - Cash Purchase → Cash refunded by cashier.
        - Store Credit/Gift Card → Credit issued.
    - Finance logs the refund in the system.
5. Customer Notification & Case Closure (Customer Service)
    - The system generates a refund receipt.
    - Store staff provides confirmation to the customer.
    - Case is marked as resolved.

Key BPMN Modeling Notes:

- External Pool:
  - Customer

- Swimlanes:
  - Customer Service → Handles refund requests and customer communication.
  - Operations (Store Manager) → Verifies and approves product return.
  - Finance (Cashier/Payment Processing) → Executes refund transaction.

- Events & Gateways:
  - Use an exclusive gateway (XOR) after the eligibility check to split between approved refunds and rejected requests.

- End Events:
  - Refund Completed (for successful transactions).
  - Request Rejected (for ineligible refunds).

📌 To obtain the End-to-End L3 BPMN model for this process, please email to kevin.tan@process-bility.com with the subject line 'Process-bility Coles'.

Process Guardian in Process-bility

In Process-bility, the traditional concept of a Process Owner is replaced with a more practical and collaborative role—the Process Guardian. Instead of being an authoritative figure in management, the Process Guardian acts as a facilitator and coordinator, ensuring that processes are well-documented, kept up to date, and effectively used by both business and technical stakeholders. They play a crucial role in cross-functional collaboration by working with, not above, the managers of different silos.

Why Process Guardian Instead of Process Owner?

The traditional role of a Process Owner often creates unnecessary tension, confusion, and inefficiencies because it implies control over processes that span multiple teams. In contrast, the Process Guardian is designed to work with, not above, functional managers, helping them align and coordinate their processes without disrupting their authority.

Key Responsibilities of a Process Guardian:

- Ensuring Process Documentation is Up to Date: The Process Guardian maintains process accuracy and ensures it reflects current business operations.
- Supporting Business & Technical Stakeholders: Helping teams use processes effectively for improvements, training, and audits.
- Facilitating Cross-Silo Agreement: Since L3 processes span multiple functional areas, the Process Guardian ensures agreement among silo managers before any changes are made.
- Managing Process Change Requests: When updates are needed, the Process Guardian is responsible for obtaining agreement from all impacted silo managers before approving changes to L3 processes.

Closing Thoughts

Designing an effective process structure is all about making processes work for the organization. With Process-bility, we take a hybrid approach, maintaining a stable functional hierarchy (L1–L2) while enabling flexible, cross-functional end-to-end processes (L3).

By introducing the Process Guardian, we shift away from rigid process ownership models and empower collaboration, alignment, and continuous improvement. This ensures that processes remain relevant, practical, and adaptable to business needs.

A well-designed process structure brings clarity, reduces inefficiencies, and enables teams to work together effectively. With Process-bility, you now have the tools to create scalable and sustainable process models that bridge the gap between silos and drive meaningful business outcomes.

Chapter 13

The Seven Layers of Process-bility for AI

AI agents don't just need prompts. They need a knowledge architecture.

There is a temptation, when deploying AI inside an organisation, to treat it like a very capable new employee on their first day. You hand them access to the systems. You point them at the documents. You give them a few instructions and assume they will figure out the rest.

That assumption works tolerably well for humans. Humans bring context. They ask questions. They pick up unspoken rules from the environment around them. They know, instinctively, when something feels off.
AI does not do any of that.

Without the right knowledge architecture underneath it, AI will act on what it can see — and confidently fill the gaps with what it cannot. The result is not intelligence. It is automation without judgment.
This is not a technology problem. It is a knowledge problem. And it is precisely the problem that Process-bility was built to solve.

Process-bility was designed for people. But it was waiting for AI.

The seven layers of Process-bility were not invented for the age of artificial intelligence. They grew from decades of observing what goes wrong when organisations try to document, share, and act on business knowledge.

Humans struggle with the volume. Seven layers of connected knowledge is a lot to hold in one person's head. In practice, most organisations document one or two layers well and leave the rest as tribal knowledge, buried in someone's inbox or locked inside the heads of their most experienced staff.
AI does not have that problem.

For AI, volume is not a burden. It is fuel. The more structured, connected, and complete the knowledge architecture, the more effectively AI can act within it. What has always been the ideal for human communication becomes, for AI, the minimum viable input.
Process-bility does not just describe how a process works. It describes the full context in which that process lives. That is what makes it AI-ready.

The Seven Layers

Each layer answers a different question. Together, they form a complete picture of how work actually happens inside an organisation.

Layer 1: Process

The process layer is the instruction set. It is the BPMN hierarchy — from the L1 enterprise view down through L2, L3, and L4 — that maps what happens, in what sequence, under what conditions, and who is responsible at each step.
Without this layer, AI has no workflow. It cannot understand that certain activities must happen before others, that some paths are conditional, or that a process has a defined start and a defined end. The process layer is not optional. It is the spine of everything else.

Layer 2: Systems

The systems layer defines the environment in which the process operates. Which applications are involved? Which APIs are available? Where does the agent need to read from, write to, or trigger an action?
For a human, system knowledge is partly habit — they know which application to open because they have opened it a thousand times before. For AI, that knowledge must be explicit. The systems layer tells the agent not just what to do, but where to do it.

Layer 3: Data

The data layer defines the raw material of the process. What data objects are created, consumed, validated, and updated at each step? What are the field definitions, acceptable values, and data relationships that govern those objects?
This layer is often the most underestimated. Organisations that have invested in systems architecture frequently discover they have built those systems on poorly defined or inconsistently used data. AI will surface those inconsistencies fast. A well-defined data layer is not just useful for AI — it is a quality test for the entire knowledge architecture.

Layer 4: Business Rules

The business rules layer captures the logic behind every decision in the process. Gateway conditions, approval thresholds, eligibility criteria, calculation methods, and exception handling all live here.

Business rules are the difference between a process that follows procedure and one that follows intent. Many organisations have rules that exist only in the heads of experienced staff, applied inconsistently and rarely written down. For AI to act on behalf of an organisation, those rules must be explicit, structured, and complete.

Layer 5: Human Element

The human element layer is the most difficult to document and the most important to understand.

Every process, as it exists on paper, differs from the process as it is actually performed. People have workarounds. They know which steps are genuinely critical and which are formalities. They understand the exceptions that the documentation does not cover. They carry the institutional memory that explains why certain things are done in ways that no longer make obvious sense.

This layer is not about romanticising human judgment. It is about acknowledging that AI trained only on official documentation will encounter the reality of how work actually happens and may not know what to do with it. The human element layer bridges the gap between the documented process and the lived process.

Layer 6: Governance

The governance layer defines the guardrails. Risk obligations, compliance requirements, regulatory constraints, audit controls, and organisational policies all belong here.

For AI, governance must be encoded as constraints on behaviour. But governance is not only about what the agent can and cannot do. It also defines when the agent should stop and hand control back to a human.

Not every decision should be made by AI. Some decisions carry risk that requires human judgment. Others have regulatory or ethical weight that demands human accountability. The governance layer is where those boundaries are drawn — the conditions under which the agent pauses, escalates, and waits.

This is the human-in-the-loop by design, not by accident. When something falls outside the agent's authorised boundaries, the governance layer tells it to step aside. When something goes wrong — and something will always eventually go wrong — the governance layer is where you look to understand what the agent was and was not authorised to do, and whose judgment should have taken over.

Layer 7: Organisational Context

The organisational context layer is the widest lens. It captures the structure of the organisation, the dependencies between teams and functions, the ownership of processes and systems, and the strategic objectives that give all of the above their purpose and direction.

An AI agent operating without this layer can complete a task correctly in isolation and still cause damage at the organisational level — because it does not understand how its actions connect to decisions made elsewhere, or why certain constraints exist beyond what the documentation says.

Context is not decoration. It is the difference between an agent that executes and one that understands.

Not just for AI agents

It is worth being clear about something. The seven layers are not a framework for AI deployment. They are a framework for organisational knowledge.

The fact that they also serve as the foundation for effective AI is not a coincidence. It is a reflection of a deeper truth: the knowledge architecture that allows AI to act with clarity and control is the same knowledge architecture that allows humans to do the same. Organisations that have always struggled to document more than one or two layers well have simply never felt the full cost of that gap — because humans fill in what is missing. AI cannot fill in what is missing. It can only work with what it is given.

This is why the arrival of AI in the enterprise is, among other things, a forcing function for process rigour. It will not be enough to have a process map on the wall and a policy document on the intranet. Organisations that want to deploy AI safely and effectively will need to invest in the full depth of their knowledge architecture.

Process-bility provides that architecture. Not as a new thing, invented for the AI age, but as a proven framework that the AI age is finally making indispensable.

The equation

Process + Systems + Data + Business Rules + Human Element + Governance + Organisational Context

That is not just a list.

That is what it means to truly know your process.

And in the age of AI, knowing your process is no longer a competitive advantage.

It is the price of admission.

Afterword

Thank you for taking the time to explore BPMN Made Easy with Process-bility.
Process-bility is not just a guide to process modeling—it's a reflection of the lessons, challenges, and aha moments I've experienced throughout my journey in BPMN modeling. My hope is that the insights, techniques, and examples shared here will inspire you to approach process modeling not just as a technical exercise but as a powerful communication tool.

In today's TikTok world, clarity and simplicity are more valuable than ever. Processes evolve, adapt, and transform, and Process-bility provides a framework to make sure you can grow with them. This book isn't just about creating models—it's about aligning people, solving problems, and driving action.

If there's one thing I've learned over the years, it's that the real magic happens when we rediscover our growth mindset—embracing new ways of thinking, collaborating with openness, and, above all, having fun along the way. So, as you go forward, I encourage you to embrace Process-bility, learn from it, and make it your own.

Thank you again for being part of this journey. Here's to turning complexity into clarity and making BPMN easy for everyone.

Warm regards,

Kevin Tan
P.S. A good process model is the one you can still understand easily after three glasses of wine.

Thank You for Reading!

I appreciate you taking the time to read BPMN Made Easy with Process-bility.
I hope this book has helped you see BPMN in a new way—with clarity, simplicity, and effectiveness.

📌 If you found this book valuable, I'd love to hear your feedback!
Please consider leaving a review on Amazon—your insights will help others benefit from Process-bility.

🌐 Stay connected for more BPMN insights:

Website: www.process-bility.com
LinkedIn: linkedin.com/in/kevin-tan-process-bility/

Process-bility Press
"Now Everyone Can Understand BPMN."

Made in United States
Orlando, FL
05 November 2026

90200400R10098